AIMING FOR AN A IN A-LEVEL BIOLOGY

Jo Ormisher

HODDER
EDUCATION
AN HACHETTE UK COMPANY

The Publishers would like to thank the following for permission to reproduce copyright material.

Photo credit

P.60, M.I.Walker/Science Photo Library

Acknowledgements

With thanks to the CDARE team at the Sheffield Institute of Education for their assistance in developing and reviewing this title.

Every effort has been made to trace all copyright holders, but if any have been inadvertently overlooked, the Publishers will be pleased to make the necessary arrangements at the first opportunity.

Although every effort has been made to ensure that website addresses are correct at time of going to press, Hodder Education cannot be held responsible for the content of any website mentioned in this book. It is sometimes possible to find a relocated web page by typing in the address of the home page for a website in the URL window of your browser.

Hachette UK's policy is to use papers that are natural, renewable and recyclable products and made from wood grown in sustainable forests. The logging and manufacturing processes are expected to conform to the environmental regulations of the country of origin.

Orders: please contact Bookpoint Ltd, 130 Park Drive, Milton Park, Abingdon, Oxon OX14 4SE. Telephone: (44) 01235 827827. Fax: (44) 01235 400401. Email: education@bookpoint.co.uk. Lines are open from 9 a.m. to 5 p.m., Monday to Saturday, with a 24-hour message answering service. You can also order through our website: www.hoddereducation.co.uk

ISBN: 978 1 5104 2409 8

Typeset in Integra Software Services Pvt. Ltd., Pondicherry, India

Printed in Spain

A catalogue record for this title is available from the British Library.

Contents

Getting the most from this book

Aiming for an A is designed to help you master the skills you need to achieve the highest grades. The following features will help you get the most from this book.

Learning objectives

> A summary of the skills that will be covered in the chapter.

✓ **Exam tip**

Practical advice about how to apply your skills to the exam.

Activity

An opportunity to test your skills with practical activities.

! Common pitfall

Problem areas where candidates often miss out on marks.

The difference between...

Key concepts differentiated and explained.

Annotated example

Exemplar answers with commentary showing how to achieve top grades.

Worked example

Step-by-step examples to help you master the skills needed for top grades.

Take it further

Suggestions for further reading or activities that will stretch your thinking.

You should know

> A summary of key points to take away from the chapter.

About this book

The A-grade student

Only about 9% of biology students achieve an A* and 18% a grade A at A-level. How do they achieve this?

This book outlines some of the behaviours demonstrated by high-achieving students and introduces the key skills that you need to develop throughout your A-level course. Each chapter includes suggestions for activities that will complement the work you are doing in class. You should not try to work through the whole book in one go, but should 'dip in' as and when you need to.

Using this book

Each chapter in this book focuses on a different skill. You will not read the chapter and perfect the skill immediately, but will need to keep revisiting each skill so that you become more competent and confident throughout your A-level course.

Chapter 1 focuses on **quantitative skills**, including handling and interpreting data. Being able to analyse and critically evaluate data is a very useful skill to develop, not just to answer exam questions, but also to determine the quality and relevance of experimental data and to make judgements about data presented in the media.

In the high-pressure exam environment, misreading or misinterpreting exam questions can cost valuable marks. Following the advice and completing the activities on **reading skills** in chapter 2 will help you to identify what a question requires of you, so you can write a suitable answer. There is no specific requirement by most exam boards for further reading, but A-grade students are likely to apply to prestigious universities and for competitive courses, so further reading is important for both your UCAS personal statement and university interviews.

All A-level exams will include a mixture of question types, ranging from those requiring one-word answers to extended writing; for some exam boards this might include writing an essay. Activities in chapter 3 on **writing skills** will help you write clear, concise answers within the time constraints of an exam.

Chapter 4 on **practical skills** includes guidance and activities to help you get the most from your practical work, with advice on the practical endorsement, written papers and practical exams.

In general, an A-grade student will spend more hours per week on independent study and will use more different learning strategies than a B-grade student. Chapter 5 focuses on **study and revision skills** and suggests activities that will develop your higher-order thinking skills, while encouraging you to analyse, justify and evaluate rather than focusing on straightforward recall of knowledge.

Chapter 6 (**exam board focus**) identifies the command words used by the different exam boards to assess the higher-order thinking skills and, most importantly, outlines key advice for students based on examiners' reports. Each exam paper includes questions that allow students to demonstrate that they have met the three assessment objectives (AO — Table 1). AO3 exam questions require the higher-order thinking skills characteristic of an A-grade student and may be used as discriminators between mark bands.

Table 1 Assessment objectives

Assessment objective	Weighting	Typical exam question
AO1 — demonstrate knowledge and understanding of scientific ideas, processes, techniques and procedures	35–40% of an AS paper and 30–35% of an A-level paper	Exam questions of this type test core thinking skills and will have command words such as 'Describe', 'Give' and 'Name'.
AO2 — apply knowledge and understanding of scientific ideas, processes, techniques and procedures in both theoretical and practical contexts, and when handling qualitative or quantitative data	40–45% of both AS and A-level papers	Applying knowledge is a higher-order thinking skill, so questions may include unfamiliar examples with command words like 'Explain' or 'Suggest'.
AO3 — analyse, interpret and evaluate scientific information, ideas and evidence	20–25% of an AS paper and 25–30% of an A-level paper	Questions may require you to make judgements, reach conclusions or refine practical design. You may be given a set of results and asked to explain them, or you may have to critically evaluate data to decide whether or not they support a given conclusion. Command words used for AO3 exam questions may include 'Evaluate' or 'Justify'.

Top tips for success

→ Focus on the future. Is there a reason you want to get a grade A? Maybe it is for a degree course, a specific university or a career you have in mind. Having a goal and a reason to work hard will help to motivate you when you feel like giving up.

→ It is OK to get things wrong. You will make mistakes, and you may not be an A-grade student at the start of the course, so try to turn your mistakes into positive learning experiences. What do you need to do to get it right next time?

→ Work hard. Nobody gets an A in biology A-level without working hard. There is a lot of content to learn and a range of skills to develop. Time and effort are essential for success.

→ Develop your exam technique. Exam questions should not be left until the end of the course. Start using them straight away so that you are familiar with the question format and mark schemes.

→ Include variety in your study. Move out of your comfort zone and try different activities, especially the more challenging ones. There are lots of activities in this book to get you started.

→ Make good use of your teachers! They are well educated, knowledgeable and experienced in coaching students to exam success. Act on their feedback, engage in their lessons and ask them for extra work.

→ Finally, but most importantly, enjoy the subject. You have chosen to study biology as one of your top three subjects. There may be certain aspects that you enjoy more than others, so find out what really sparks your interest. From a bacterium to a blue whale, there is a beautiful and varied world of living things for you to explore.

1 Quantitative skills

Quantitative skills require the use of numerical data in a variety of different formats. Developing these skills is essential for the successful biologist as they underpin all experimental work, including:

→ devising methods using **sampling** techniques and a suitable **range** for variables
→ recording data using suitable **units** and to an appropriate number of **decimal places**
→ displaying data in **bar charts**, **histograms** or **scatter diagrams**
→ analysing data using **statistical tests**
→ evaluating scientific procedures, results and conclusions

Core study skills

Knowledge and understanding

The assessment of quantitative skills in biology A-level will include at least 10% Level-2 or above mathematical skills. Core mathematical skills include simple substitutions, with little choice of data or formulae, whereas Level-2 mathematical skills require:

→ application and understanding, with a choice of data or formulae
→ problem-solving skills
→ A-level mathematical content, for example exponential and logarithmic functions

The most challenging calculation questions will have multiple steps and require the understanding of several mathematical skills. The following worked examples suggest methods of approaching some of the questions aimed at the A grade student.

Worked example 1.1

This example includes understanding of units and percentages as well as numerical computation.

A fish can extract 90% of the oxygen from water as the water flows over its gills.

The oxygen concentration of water at 20°C is 9 mg dm^{-3}.

A 0.3 kg fish has an oxygen consumption at rest of 300 mg kg^{-1} hour^{-1}.

Calculate the volume of water in dm^3 that needs to pass over the gills of the fish every hour to supply the oxygen required. (2)

Step 1: Find the oxygen consumption of the fish in the question.

The oxygen consumption at rest is 300 mg kg^{-1} hour^{-1}, but the fish in the question is only 0.3 kg.

The oxygen consumption of the fish is $300 \times 0.3 = 90$ mg hour^{-1}.

Step 2: Find the mass of oxygen that the fish can extract from 1 dm^3 of water.

Water has an oxygen concentration of 9 mg dm^{-3}, but the fish can only extract 90% of the oxygen:

90% of 9 mg dm$^{-3} = 0.9 \times 9 = 8.1$ mg dm^{-3}

Step 3: Calculate the volume of water that needs to pass over the gills every hour.

Divide the hourly oxygen consumption of the fish by the mass of oxygen in the water:

$$\frac{90\,mg}{8.1\,mg\,dm^{-3}} = 11.1$$

The volume of water that needs to pass over the gills every hour is 11.1 dm^3.

 Exam tip

As well as past papers and mark schemes, you can also read the 'Examiners' report' for each exam. This includes a general comment about students' overall performance on the paper as well as specific comments about each question. The report provides a useful insight into common misconceptions or errors that have led to marks being lost.

 Exam tip

When faced with a challenging question, just break it down into simple steps and show your working clearly. You may get credit for intermediate steps, even if the final answer is incorrect.

Worked example 1.2

Some exam questions require you to manipulate data *and* use the data to support conclusions.

Scientists investigated the arrangement of phospholipid molecules in the cell surface membrane of red blood cells.

The scientists found that:
- **the number of red blood cells in 1 mm^3 of blood = 4.37×10^6**
- **the mean surface area of the red blood cells = 152 μm^2**
- **the surface area of the phospholipids extracted from 1 mm^3 of blood = 13.1 cm^2**

The scientists concluded that the phospholipid molecules are arranged in a double layer.

How do the data support this conclusion? (3)

This is another challenging question due to the number of steps involved.

You need to be confident with the use of numbers in standard form, and also with conversion between units. Check that you can convert between μm² and m².

One way of answering this question might be:

Step 1: Calculate the total surface area of the red blood cells in 1 mm³ of blood.

total surface area = $(4.37 \times 10^6) \times 152 \, \mu m^2 = 664\,240\,000$ or $6.6424 \times 10^8 \, \mu m^2$

Step 2: Convert the total surface area from μm² to cm² (to compare surface area of red blood cells with surface area of phospholipids).

$1 \, \mu m = 1 \times 10^{-4} \, cm$

$1 \, \mu m^2 = 1 \times 10^{-8} \, cm^2$

total surface area of red blood cells = $6.6424 \times 10^8 \, \mu m^2 = 6.6424 \, cm^2$

Step 3: Compare the total surface area of the red blood cells with the surface area of the phospholipids.

total surface area of red blood cells = $6.6424 \, cm^2$

Doubling the surface area of the red blood cells gives $13.28 \, cm^2$, which is similar to the value of $13.1 \, cm^2$ for the surface area of the phospholipids.

The data do support the conclusion because the evidence suggests that phospholipid molecules are arranged in a double layer.

Worked example 1.3

You may be provided with data about unfamiliar organisms and asked to apply your mathematical skills to these data.

Llamas are mammals that live at high altitudes. Llama haemoglobin differs from human haemoglobin. Figure 1.1 shows the percentage saturations of llama and human haemoglobin at different partial pressures of oxygen.

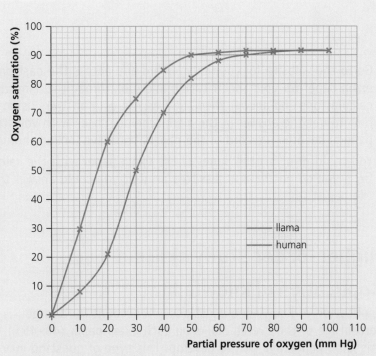

Figure 1.1 Oxygen dissociation curves for human and llama haemoglobin

➥

(a) Draw a tangent on each curve to find the rate of change in oxygen saturation for both llama and human haemoglobin at 45 mmHg. (2)

(b) Find the ratio of rate of change for human haemoglobin : llama haemoglobin. (1)

ratio = _____ : 1

This is another question with multiple steps. Remember to:
- draw the tangents so that they *just* touch the lines at 45 mmHg, making sure that you use a sharp pencil
- express the ratio in the order specified by the question, with the rate for human haemoglobin given first

Step 1: Find the rate of change in oxygen saturation for llama haemoglobin at 45 mmHg.
- Draw a tangent on the curve for llama haemoglobin that just touches the curve at 45 mmHg.
- Draw a triangle with the tangent as the hypotenuse.
- Read off the values for the height of the triangle from the y-axis to find the change in y.
- Read off the values for the width of the triangle from the x-axis to find the change in x.
- Calculate the rate of change by diving change in y by change in x:

$rate\ of\ change = 0.5\%\ mmHg^{-1}$

Step 2: Find the rate of change in oxygen saturation for human haemoglobin at 45 mmHg

Repeat step 1 using the curve for human haemoglobin:

$rate\ of\ change = 1.2\%\ mmHg^{-1}$

Step 3: Find the ratio of change.

Divide the rate for human haemoglobin by the rate for llama haemoglobin:

$ratio = 2.4 : 1$

Activity

Go to the website for your exam board and access the past papers section. Select the most recent exam paper for your specification and look through it. Find the question parts that involve calculations and answer them. Self-assess your answers using the mark scheme. If you are unsure how to get the answer in the mark scheme, check with your teacher. Repeat this with other past papers, keeping a record of the papers you have looked through.

You could print off the pages of the exam paper with the calculations and keep them in a folder. Or, if you do not have access to a printer, you could answer the questions in an exercise book or on file paper.

Higher-order study skills

Application of knowledge

Application of knowledge questions will account for a large percentage of the overall marks on each exam paper (40–45%), so this is an important skill to develop. This type of question may require you to look at quantitative information and explain it using your scientific knowledge and understanding.

Quantitative information could be presented in a number of different ways, including:

→ tables
→ line graphs
→ bar charts
→ scatter diagrams

There are some golden rules for answering this type of question:

1 Identify the main topic of the question — for example, enzymes or respiration.
2 Look for patterns or trends in the data.
3 Think of reasons *why* these patterns or trends have occurred.

Worked example 1.4

An exam question assessing higher-order skills may contain a lot of unfamiliar introductory information in the stem of the question, or data for you to interpret.

Yeast are eukaryotic, single-celled microorganisms. They can respire both aerobically and anaerobically. Anaerobic respiration in yeast produces carbon dioxide and ethanol. Ethanol is toxic to yeast at high concentrations.

A student investigated respiration in yeast. The student:

● **added yeast to a liquid growth medium containing glucose**
● **cultured the yeast in a sealed conical flask for 24 hours**
● **measured oxygen uptake and ethanol production every hour**

The student's results are shown in Figure 1.2.

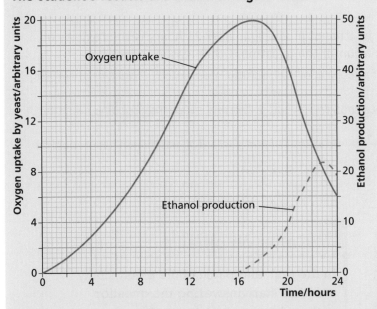

Figure 1.2 A graph to show the effect of time on oxygen concentration and ethanol production in yeast

Describe and explain the student's results. (4)

The main topic of this question is respiration, so even if you are unfamiliar with yeast, you can still apply your knowledge and understanding of the topic.

Describing the results requires you to state any overall patterns or trends, and to use relevant data from the graph. This graph has two *y*-axis scales: the *y*-axis on the left is for the oxygen uptake and the *y*-axis on the right is for ethanol production.

Step 1: Describe the oxygen uptake over time.

Oxygen uptake rises steadily between 0 and 17 hours, reaching a maximum of 20 arbitrary units. After 17 hours, the oxygen uptake decreases rapidly.

Step 2: Describe the ethanol production over time.

Ethanol production begins at 16 hours and rises steadily between 16 and 23 hours, reaching a maximum of 22 arbitrary units at 23 hours. After 23 hours, ethanol production decreases.

Explaining the results requires you to apply your scientific knowledge to this unfamiliar example and give reasons *why* the changes in oxygen uptake and ethanol production might have occurred.

Step 3: Explain the shape of the oxygen uptake curve.

Structure your answer logically and firstly explain why oxygen uptake increases, then why it decreases.

Uptake increases because oxygen is required for aerobic respiration. The yeast cells will divide during the investigation, so more yeast cells will be present and this will increase the demand for oxygen.

The yeast is cultured in a sealed conical flask. Without the addition of oxygen to the culture, there will be less oxygen available and the oxygen uptake will decrease.

Step 4: Explain the shape of the ethanol production curve.

Ethanol production does not begin until 16 hours. You are told that anaerobic respiration in yeast produces ethanol, so the yeast is only respiring aerobically between 0 and 16 hours, and then begins to respire anaerobically at 16 hours.

One possible explanation for the decrease in ethanol production after 23 hours is the toxicity of ethanol to yeast. You are given this information in the question. Other reasons might be:
- lack of glucose — *without the addition of a respiratory substrate, respiration (aerobic or anaerobic) cannot continue*
- lack of space — *the yeast will divide throughout the investigation and space may become a limiting factor*
- accumulation of other toxic substances — *by-products of metabolism other than ethanol could be toxic to the yeast*

! Common pitfall

Students often waste time *explaining* results when they have just been asked to *describe* them.

! Common pitfall

In the rush and panic of the exam, many students start answering the question without having read it properly.

Analysis, interpretation and evaluation

Exam questions requiring analysis, interpretation or evaluation can be the most challenging questions on the exam paper. To answer them successfully you need the background knowledge and understanding of the topic, you need to be able to apply this knowledge to an unfamiliar situation, and then you have to **analyse**, **interpret** or **evaluate** information.

Activity

Go to the website for your exam board and search for the command words. Make a glossary of the command words and what they mean. (Good students often waste time by writing too much for each question instead of focusing on the command word and doing exactly what the examiner wants them to do.) Find a past paper question with quantitative data — a table, a graph or both. Look at the exam questions about the table or graph and then try substituting different command words. For example, if a question asks you to **describe** the results, how would you answer the question if you were asked to **explain** the results, or **evaluate** the results?

Analysing and interpreting

'Analyse' as a command word means 'separate information into components and identify their characteristics'. For example, you could be given data in a table or a graph and you would need to **analyse** the data and identify patterns.

You should be able to make sense of the data by looking at the range, calculating means, rates and standard deviations, or determining whether the standard deviations overlap.

Interpret means that you have to 'translate information into recognisable form'. Having analysed the data, you need to interpret the findings. For example, you could be given information in the form of graphs or charts and would need to use this data to draw conclusions.

You should be able to **interpret** the results of statistical tests. Many students write answers that indicate a fundamental misunderstanding of statistical analysis. Statistical concepts are often incorrectly or poorly expressed — for example, stating that 'the results are significant' or '5% of the results are due to chance'. Answers should state that 'the *difference between* the means is significant' rather than the suggesting that the means or data themselves are significant. Non-significance does not mean that the data have been collected incorrectly, or that the method was wrong.

You also need to **interpret** graphs — for example, showing pressure change in the heart, or oxygen dissociation curves, and make links to biological information.

Worked example 1.5

This question gives you a complicated-looking graph that you need to interpret so that you can explain what the graph shows.

The cardiac cycle is the sequence of events that take place during one heartbeat. Blood flows through the heart due to differences in pressure. Figure 1.3 shows the pressure changes in the left side of the heart during one heartbeat.

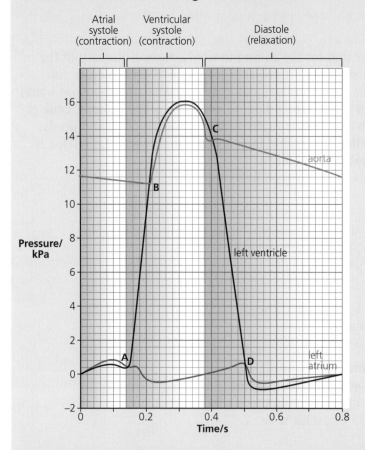

Figure 1.3 Pressure changes in the aorta, left ventricle and left atrium during one cardiac cycle

Describe and explain the changes to the atrioventricular or semilunar valves at points A, B, C and D on the graph.   **(4)**

Step 1: Identify the pressure changes that are happening at points A, B, C and D on the graph.

At point A, the pressure in the left ventricle increases and becomes greater than the pressure in the left atrium. Annotate the graph to show this with LV > LA. Repeat this with points B, C and D on the graph.

Step 2: Apply your knowledge of heart structure and valve function to interpret the graph.

Remember that blood moves from high pressure to low pressure. You need to visualise the structure of the heart and apply your knowledge of valve function to explain what happens at each point on the graph.
- Higher pressure in the atrium than the ventricle will *open* the atrioventricular valve.
- Higher pressure in the ventricle than the atrium will *close* the atrioventricular valve.
- Higher pressure in the left ventricle than the aorta will *open* the aortic semilunar valve.
- Higher pressure in the aorta than the left ventricle will *close* the aortic semilunar valve.

Step 3: Write four sentences to describe and explain the changes to heart valves.

Each sentence should clearly refer to the letter on the graph, the opening or closing of a named valve, and an explanation of the pressure difference that causes this opening or closing.

At point A, the pressure in the left ventricle becomes higher than the pressure in the left atrium, causing the atrioventricular valve (between the left atrium and left ventricle) to close.

The same principle of pressure difference can be applied to valves opening or closing at points B, C and D.

Evaluating

Evaluating is a higher-order skill and requires you to look at scientific procedures, data and conclusions from a critical viewpoint. An 'evaluate' question requires a balanced answer, giving pros and cons, or evidence for and evidence against. You should analyse the available evidence and make a judgement. This skill is not just useful when answering exam questions, it is essential in everyday life to make lifestyle and healthcare choices, and to make sense of the stories reported in the media.

You could present your arguments in favour using phrases like:
→ Evidence for the conclusion...
→ Yes, the data do support the conclusion because...
→ Yes, the scientist was correct because...

Use connectives to present the other side of the argument, for example:
→ Alternatively...
→ However...
→ On the other hand...
→ Whereas...

Finish by writing a concluding sentence:
→ In conclusion...
→ In summary...
→ On the whole...
→ To conclude...

Evaluating scientific procedures

Exam questions may describe an investigation that has been carried out by a scientist or by a student. As you read through the procedure, you should consider the following factors (Table 1.1):
→ Controls
→ Duration
→ Sample

Exam tip

Exam papers usually become progressively more difficult as you work your way through. The last one or two questions on a paper are likely to be the ones aimed at differentiating between the A and A* students.

Table 1.1 Evaluating scientific procedures

Factor	Evaluation	Explanation
Controls	Has a **control experiment** been used?	A control experiment is set up to eliminate the possibility that something other than the independent variable may have caused the results.
	If a test group has been given a drug, what has the **control group** been given?	Control groups should be treated in exactly the same way as the test group, but should not be given the drug. The test group may be given a **placebo**, or they may continue to take their original medication. For example, if testing a new medication for the treatment of asthma, the control group would continue to use their normal medication, as it would be dangerous for them to discontinue this and be given a placebo.
	Which **variables** were **controlled** and how were they controlled?	Control variables can affect the outcome of the investigation and so must be kept constant or monitored. If testing a drug, controls are necessary to show that it is the drug that is responsible for any difference between the two groups.
Duration	How long was the study carried out for?	Investigating the effect of smoking on lung cancer would show no effect if the smokers were only investigated over a short period of time. However, an investigation over many years would show the link between smoking and cancer.
Sample	What was the sample size?	The larger the sample size, the more representative of the population as a whole.
	How have test subjects been selected or assigned to groups?	Random sampling avoids bias.
	Who, or what, has been tested?	Has a trial used isolated cells, human subjects or a different animal like a rat? Results obtained using isolated cells or different animals cannot always be replicated in humans.

Activity

Look through a newspaper or magazine for a story or advert with scientific claims. For example, adverts for beauty products often refer to the percentage of people reporting an improvement in hair thickness or wrinkle depth. Newspapers often have new health claims about foods or supplements. Read these articles from a critical point of view:

- How big was the sample size?
- How long did the study last?
- Has the substance been tested on humans?
- Who funded the study?

Evaluating data

Data could be presented in tables, graphs and charts. Tabulated data are often displayed using graphs and charts because this makes it easier to spot patterns and trends. Data can be manipulated to show what we want them to show, so it is important to look at the data critically and evaluate them. When evaluating data, look carefully at:

→ *p* values — have *p* values been given for the results? The *p* value is the probability that the differences or associations in the results are due to chance. If the probability is less than or equal to 5% ($p \leq 0.05$), then the differences or associations *are* significant (unlikely to be due to chance).

→ range — have range bars been included on a graph? The range gives the minimum and maximum values in a data set.

→ repeat measurements — have the measurements been repeated to allow a mean to be calculated and anomalous results to be identified?

→ scales — look at the scale divisions. How big are the increments? How would the conclusions be affected if a different scale had been used?

→ standard deviation — has the standard deviation been given for the data? Or have error bars been included on a graph? Error bars using standard deviation are better than range bars as they reduce the effect of extreme values. A large standard deviation shows a large spread around the mean. If the standard deviations overlap then there is probably no significant difference between the two sets of data. CCEA distinguishes between the standard deviation of the sample as an estimate for the population and the standard deviation of the mean. The error bars plotted on graphs will be the 95% confidence intervals.

→ units – have the numbers been given per thousand or per million? What unit has been used?

> ✓ **Exam tip**
>
> If you had two mean results of 92.3% and 95.4%, you could display these results as a bar chart with a *y*-axis scale from 92.0–96.0%, or you could have a *y*-axis scale from 0–100% The first scale would show a big difference between the two mean values, whereas the second scale would show very little difference between the mean values.

Take it further

Read the work of Dr Ben Goldacre, who writes about the misuse of science and statistics. He has published books, written a column for the *Guardian* newspaper and has a website from which you can access archived articles.

Evaluating conclusions

You may be given a conclusion and ask to evaluate it. For example, a question with the phrasing 'Do these data support this conclusion?' requires you to look at both evidence supporting the conclusion and evidence against the conclusion.

Worked example 1.6

Asthma is a disease that affects the bronchioles. During an asthma attack, the muscles in the walls of the bronchioles contract. Muscle contraction causes narrowing of the bronchioles.

Doctors investigated the effect of a drug called albuterol on the forced expiratory volume (FEV) of asthma patients. The FEV is the volume of air forced out of the lungs in the first second of exhalation.

Patients were separated into three groups:
- **Group 1 were given an inhaler containing albuterol.**
- **Group 2 were given an inhaler containing a placebo.**
- **Group 3 were given no inhaler.**

The FEV of the patients was measured at the start of the investigation and after 60 days of treatment.

The doctors calculated the mean percentage increase in FEV for each group. The doctors' results are shown in Figure 1.4.

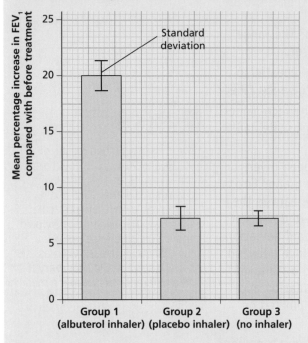

Figure 1.4 A bar chart to show the effect of albuterol spray on mean percentage increase in FEV compared with control groups

The doctors concluded that albuterol is the most effective treatment for asthma.

Evaluate this conclusion.   (3)

Step 1: Evaluate the procedure.

Pros of the procedure include the use of control groups, testing on humans and the relatively long duration of the study. The cons include the lack of information about the number of patients, no indication of the severity of the disease, or of how patients were allocated to each group.

Step 2: Evaluate the results.

Pros of the results include an obvious difference between the test group and the control groups, the calculation of standard deviation and the lack of overlap of error bars (which indicates that the difference may be significant). The cons include no evidence of repeat measurements, a lack of statistical analysis and no p value to determine whether the differences between the results are significant.

Step 3: Evaluate the conclusion.

Evidence for the conclusion includes the larger percentage increase in FEV with albuterol spray (+20%) compared with without albuterol spray (+7%). However, the procedure and results both have omissions that mean that the conclusion may not be valid. In addition, there is no explanation for the increase in FEV in the control groups.

Application to the exam

Sample question

This exam question assesses a number of the skills discussed in this chapter. There is guidance on how to approach the question, a mark scheme so you can self-assess your answer, and then a sample answer at the end.

Glioblastoma is a type of brain tumour.

Radiotherapy has been used to treat glioblastoma.

Scientists investigated the use of a new drug, temozolomide (TMZ), to treat glioblastoma.

573 patients were randomly assigned to one of two groups:
➡ **Group A were just treated with radiotherapy.**
➡ **Group B were treated with radiotherapy and TMZ.**

Doctors compared the median survival time for each group. The survival time is how long a patient lives following diagnosis.

The doctors also compared the 2-year survival rate.

Table 1.2 shows the doctors' results.

Table 1.2 The effect of different treatments for glioblastoma on the median survival time and the percentage survival at 2 years

Treatment	Median survival time/months	Percentage of patients alive after 2 years
RT alone	15	20
RT/TMZ	21	43

The doctors carried out a statistical test and obtained a *p* value of 0.006.

(a) **How would the doctors find the median survival time?** (2)

(b) **Group A was the control group. Why was this group treated with radiotherapy instead of being given a placebo?** (1)

(c) **Explain what the results of the statistical test show.** (2)

(d) **The doctors concluded that TMZ improved the treatment of glioblastoma. Do the data in Table 1.2 support this conclusion? Give reasons for your answer.** (4)

How to approach the question

Read through the information slowly and think about each statement. Write additional notes next to the question if it helps your understanding:

➡ You may not have heard of glioblastoma, but you may have studied cancer as part of the A-level course.

➡ 573 patients is a large sample size. A sample size needs to be large enough to carry out statistical analysis.

➡ 'Patients' tells you that the investigation was on humans. Sometimes you are given the results of trials on other animals and so the drug being investigated may not have the same effect in humans.

→ The patients are *randomly* assigned to one of two groups. Randomisation prevents selection bias, ensures that each patient has an equal chance of receiving the experimental treatment, and produces comparable groups.

Could other factors affect the outcome of the investigation? For example:

→ the age of the patients

→ gender

→ the size of the tumour

→ secondary diseases like high blood pressure or diabetes

→ lifestyle factors such as drinking or smoking

Be *critical* when looking at results. Think about what you have *not* been told. For example:

→ How many are in each group?

→ What are the upper and lower values for survival times?

→ What is the quality of life?

→ Are there any side-effects of the drug?

Give yourself 15 minutes to answer the question.

Mark scheme

(a) 1   arrange survival times in order from low to high (or high to low);

 2   find the value with same number of survival times above and below/find the middle value;

(b) (patients in the control group had brain tumours so) unethical not to treat/should still receive best treatment currently available;

(c) 1 p value suggests that difference is significant;

 2   difference between the results of the two groups is unlikely to be due to chance;

(d) Yes, the results do support the conclusion because:

 1   median survival time increases by 6 months;

 2   number of patients still alive after 2 years is 23% higher;

 No, the results do not support the conclusion because:

 3   no standard deviation given and SD may overlap;

 4   some patients may have no increase in survival time;

 5   quality of life has not been reported;

 (4 max)

> **! Common pitfall**
>
> Students often state 'the results are due to chance' rather than 'the difference between the results is due to chance'.

Sample answer for (d)

Question parts (a)–(c) are very straightforward to mark, but a sample answer is included here for part (d).

Yes the data do support the conclusion because they show an increase in median survival time of 6 months ✓ (from 15 to 21) and the 2-year survival rate has more than doubled from 20% to 43% ✓. However, this means that the majority of patients (57%) did not survive for 2 years. Also, no mean survival times and standard deviations have been given. If the standard deviations overlapped, then there may be no significant difference between the mean survival times. ✓ The median survival time is given, but there could be a large range of survival times, with some patients having no increase at all. ✓ Survival time should not be the only measure of the success of a treatment. The doctors concluded that the drug 'improved the treatment', but there are no data relating to quality of life. ✓

Marking points 1 and 2 are awarded for this clear sentence that shows understanding of the data rather than simply repeating values from the table.

Marking point 3 is awarded here for a very good point about the lack of mean and standard deviation.

Good points are made about the data not supporting the conclusions, so marking points 4 and 5 have been awarded.

Overall, this is a well-balanced answer as it gives equal weighting to data supporting and not supporting the conclusion.

You should know

> Familiarise yourself with all of the mathematical skills that can be assessed at A-level *and practise them*.
> Always present your calculations clearly and show all of your working.
> Be critical when you are looking at data and think about what you have *not* been told as well as what you have.
> Write balanced arguments, with evidence in favour *and* evidence against when you are answering an evaluation question.

2 Reading skills

Learning objectives

> To use all features of your core textbook
> To develop your reading skills so you can read more efficiently
> To guide your further reading
> To develop your critical reading skills

Developing your reading skills during the 2-year A-level course will enhance your performance in the exams, but will also prepare you for study at university. At degree level you will be expected to study independently, and you will have a lot more non-contact time than you do at school or college. (Non-contact time is when you are not in lectures or tutorials, but are expected to work without direct supervision.)

There are lots of different sources available for your further reading, including:

→ textbooks
→ websites
→ journals
→ popular science books
→ scientific papers

This chapter will help you to select and use reading material, and then will focus on the higher-order skills of critical reading and evaluating source materials. The chapter finishes with advice on using your reading skills in exams, suggestions on how to apply your skills and some worked examples.

Study skills

Using core resources

Your school or college may have given you an A-level textbook for use during the course. Make sure that you use this textbook rather than letting it gather dust on a shelf for 2 years! These textbooks are reviewed and rewritten every time the specification changes, but always refer to the specification as the definitive guide to the course content.

Most textbooks have 'test yourself' questions or summary questions throughout each chapter. Answer these questions once you have read a chapter to check your knowledge and understanding. You may also find exam-style questions at the end of each chapter that you can answer and then self-assess to check that you can apply your knowledge. The answers to these questions can either be found at the back of the textbook or you may have to access them online.

The most important sections of the textbook for the A-grade student are the application and extension sections. These activities encourage you to think more deeply about the course context and practise the higher-order skills that will be assessed in the exams. Complete these activities as part of your independent study. Keep your work together in a well-organised manner so that you can refer back to it when revising:

→ Write the topic title as a main heading.

→ Include the page number and activity title.

→ Clearly number all question parts.

→ Write out any corrections in a different colour so that they stand out clearly.

Taking notes

There are different reasons why you may need to take notes while you are reading. It could be because you are using your textbook to write revision notes, or researching a specific topic using a variety of sources. But what do you need to write down and how should you write it?

The following example shows how one section from a textbook can be used to produce brief summary notes (Figure 2.1). Notice how you need to read the information and process it so that you can express it in a concise form.

> **! Common pitfall**
>
> Very able students tend to be thorough and like to write detailed notes, but there is no point in copying word-for-word from a book. Do not spend too much time writing out unnecessary information. Use the guidance in this book to help you identify the key facts and to learn how to omit words or use shorthand to produce summary notes.

Ultrafiltration

Blood enters the glomerulus from a branch of the renal artery at high pressure. Notice in Figure 2.14 that the diameter of the efferent arteriole is narrower than the diameter of the afferent arteriole; this builds up a head of pressure to force small molecules into the Bowman's capsule. This is **pressure filtration** that occurs in all capillaries but there are structural adaptations in the glomerulus to make this even more effective at removing substances from the blood. Filtration here is known as **ultrafiltration**. The hydrostatic pressure of the blood which forces fluid out of the capillaries is opposed by the oncotic pressure of the proteins in the plasma (see page 151 in the *OCR A level Biology 1 Student's Book*). The filtrate also has a hydrostatic pressure and an oncotic pressure, although both of these are very low. The net effect of these four pressures is an overall pressure forcing substances from the blood into the filtrate.

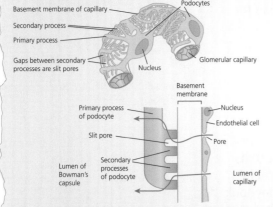

Figure 2.15 Ultrafiltration in the glomerulus. Endothelial cells forming the capillary walls have many pores in them and podocyte cells have slit pores to reduce the resistance to the flow of filtrate.

This region is well adapted for filtration because the endothelial cells lining the capillaries have pores in them that allow substances to leave the blood. On the outside of these cells is a basement membrane made of glycoproteins. This acts as a sieve retaining all the blood cells and platelets. The basement membrane allows substances with a relative molecular mass (RMM) of less than 69 000 through the glomerular capillaries into the filtrate. Most of the proteins in the plasma are larger than this so they are retained in the blood. The capillaries in the glomerulus are supported by podocyte cells that form an incomplete layer so they do not offer any resistance to the flow of filtrate.

The Kidney — Ultrafiltration

→ Blood at HP → glomerulus from branch of renal artery (diameter of **a**fferent arteriole > diameter of **e**fferent arteriole)

→ HP forces small molecules → Bowman's capsule (= **ultrafiltration**)

→ **Hydrostatic pressure** of blood → fluid forced OUT

→ **Osmotic pressure** of plasma proteins OPPOSES hydrostatic pressure

→ Filtrate has very LOW hydrostatic and osmotic pressure

→ Overall pressure forces fluid FROM blood TO filtrate

Adaptations for filtration:

1 **Endothelial cells** lining capillaries have pores — substances can leave blood

2 **Basement membrane** made of glycoproteins outside endothelial cells — retains blood cells and platelets (substances with RMM < 69 000 can pass → filtrate; most plasma proteins have > RMM so remain in blood)

3 Capillaries in glomerulus supported by podocytes — incomplete layer so no resistance to filtrate flow

Figure 2.1 Sample section from an OCR A-level Biology student textbook, with accompanying notes

Research

When researching a topic, you may need to read information from several different sources and take notes.

Make a note of the source so that you can find it again; this may be, for example, the full website URL (with date of access), or the title, author and page numbers of a book.

Do not copy out chunks of text. Read a section and then summarise it in your own words, or write down the key ideas as bullet points as you are reading. If you are quoting directly from the text, this should be made clear, for example, by using quotation marks.

The internet provides a vast source of readily available information, but you need to learn how to use it wisely. Do not assume that everything that you read is true. The section on 'Critical reading' (page 27) will help you to evaluate sources of information.

> ### Take it further
>
> The BBC News website has a science section with daily reports on new research. A particularly useful feature of these reports is the reference they make to the journal that has published the research, so you can take it further by reading the actual scientific article.

Common pitfall

It can be tempting to cut and paste information directly from a website into a Word document to save writing it out again. By doing this, not only are you not thinking about and processing the information but, more importantly, this is plagiarism — passing off someone's work or ideas as your own. All sources should be clearly cited to acknowledge the original author.

Remember that research does not just have to be using the internet. Other excellent sources of information include:
→ documentaries
→ lectures or talks
→ email correspondence or other communication with experts

Types of reading

There are different types of reading that you will need to use for different purposes. These include:
→ skimming
→ scanning
→ intensive reading
→ extensive reading

Skimming

Skim reading allows you to get the gist or general idea of a text, but with very low levels of comprehension and only a superficial understanding of the content. People often skim read as they look through newspapers so that they can identify the stories that they want to read in more detail. Skimming is useful if you want to:
→ check through a book to see if you want to read it
→ look through the options from a search engine to see which ones are relevant
→ see if an article has content relevant to your research

Scanning

Scanning is when you read very quickly to look for specific pieces of information. For example, if you were researching the importance of vaccination in the eradication of disease, you would **scan** the contents page to identify the chapter on immunity, then you would **scan** through the chapter to find the section describing the importance of vaccination. Remember to scan illustrations as well as text.

Intensive reading

Intensive reading is reading in detail with specific learning goals or tasks. You may be revising content for a test or exam, or you could be researching a topic. The key to successful intensive reading is to make sure that you are reading **actively** rather than **passively.**

Active versus passive reading

It is easy to convince yourself that you are working hard because you read your notes or textbook, but you could be reading **passively**, meaning that you are not really engaging with the text. Your eyes might be looking at the words, but are you really thinking about what they mean? Active reading is when you read with the intention of understanding and evaluating the material. As the name suggests, **active** reading involves you *doing something* while you are reading. Examples of these activities include:

→ highlighting or underlining key terms or processes
→ using different colours of pen, for example, for key terms or biological processes
→ jotting down notes, questions or ideas as you read — write in the margins, or use post-it notes
→ writing a brief 100–200-word summary of your reading at the end of an article or chapter
→ making links to other books or articles that you have read
→ working with a study partner and discussing the text

Annotated example

The text below is from page 23 of the CCEA Biology A2 Unit 1 Student Guide. The annotations suggest ways in which you might **intensively** read the text, with ideas for questions or further research.

The importance of vaccination in society

By giving **long-term immunity** to many diseases, vaccines have been important in preventing formerly common and dangerous **epidemics**. Through their use, **smallpox has been eradicated** worldwide, while measles, polio, rubella and tetanus are now **uncommon in Britain and Ireland**. Presently in the UK, vaccination offers protection from diphtheria, tetanus, whooping cough, polio, meningitis, measles,

Key terms to highlight — check that you can define them.

Why eradicated worldwide when other diseases not?

How uncommon? Why only Britain and Ireland? What about other countries?

mumps, rubella and TB. This protection relies on a high proportion of the population being vaccinated, since chains of infection are likely to be disrupted. The greater the number of individuals who are immune, the smaller the probability that those who are not immune will come into contact with an infectious individual, a situation known as **herd immunity**.

Key terms to define

How much have infant mortality rates changed?

Vaccination programmes over the past two centuries have caused **infant mortality rates** to **plummet**, allowed greater educational attainment, and improved the quality of family life and the **vibrancy of the community** in general. There are additional **economic benefits**: reduction in the medical costs of treating disease (which would otherwise be much greater than the cost of vaccination programmes) and, with fewer days off work, improved productivity.

What might the author mean by this phrase?

Are there other economic benefits not listed?

Extensive reading

Extensive reading is when you read texts for enjoyment. This type of text could be a magazine article or a popular science book, but it should be interesting and engaging. Focus on the pleasure of reading without making notes or writing questions as you go along. This type of reading is fluid and develops your vocabulary and knowledge, helping you to form your own ideas and opinions.

 Take it further

Although *The Selfish Gene* by Richard Dawkins was first published in 1976, its ideas about evolutionary biology are still relevant over 40 years later. It presents a 'gene's eye view of life', with the suggestion that animal behaviour, including altruistic behaviour, is due to the 'desire' of our genes to perpetuate.

Reading beyond the specification

You may have been told to 'read beyond the specification', but what does this mean? And what should you read? This type of reading will be **extensive** reading, where you take pleasure from the process. Take the opportunity to develop your own interests within the field of biology:

→ Find out more about the history of science. How has the work of scientists over hundreds of years led to the understanding we have today?

→ Research how theories have arisen, developed and changed, for example, what other models of membrane structure were suggested prior to the fluid-mosaic structure accepted today.

→ Look for reading lists for specific universities or as prerequisites for particular courses, like medicine or biochemistry.

→ Keep a record of what you have read, together with notes of any key facts or points you could refer back to.

→ Have an opinion about what you have read. Did you enjoy the book? If so, why did you like it? Or what did you not like about it?

→ Read books with controversial ideas or books written from different viewpoints to help you to develop your own opinions.

Take it further

On the Origin of Species by Charles Darwin is the book in which Darwin introduced his theory of evolution by natural selection. It was highly controversial at the time of its publication and is still considered to be one of the most important science books of all time. Although it is over 150 years since it was written, its format is conversational and accessible.

Critical reading

Critical reading involves the higher-level thinking skills of analysis, interpretation and evaluation. As you read a text, you should be questioning both the text and your understanding of it. The following sections suggest some of the aspects of a text that you could be considering as you read critically.

Evaluate the author

→ Who wrote it?

→ What are their credentials?

→ Are they scientists? If so, what are their qualifications and where do they work? Are they affiliated to a university? Cordelia Fine, author of *Testosterone Rex* and winner of the Royal Society Science Book of the Year 2017, is a Professor of History and the Philosophy of Science at the University of Melbourne. She holds a first class honours degree from Oxford University, an MPhil from Cambridge University and a PhD from University College London.

→ Why did they write it? Is it an angry doctor venting about a hospital that mistreated them?

→ How or what does the author stand to gain from writing the article? Was their research sponsored by a company?

Evaluate the source

→ Who owns or edits the **website**? Is it the NHS or a government?

→ Is the same story reported in different ways in different **newspapers**?

→ Is the source a published **book**? If so, it will have been edited by the publisher.

→ Has the article been published in a **scientific journal**? If so, the research will have been peer reviewed.

Activity

Use the internet to research the MMR vaccine. Find out some *factual* background information about the vaccine and its uses, including:

- the diseases the vaccine protects against
- who the vaccine is given to
- the side-effects of the vaccine
- the effects of getting the diseases

Now research the safety of the MMR vaccine and look at the websites that call for the vaccine to be banned, or link it to safety concerns. For example, you could search for MMR links to autism.

As you complete this research, *critically* read the sources, using the guidance questions in the 'Critical reading' section of this chapter.

Analyse the style of writing

→ Is it scare-mongering?

→ Is the author biased? Or is the article presenting a balanced view?

→ Is the article based on fact? Or is it the author's opinion?

Make sure that you can identify what is *fact* and what is *opinion*:

→ A fact is a statement that can be proven to be true. Facts *cannot* be changed.

→ Opinions are people's beliefs or how they feel about something. Different people have different feelings and therefore different opinions. Opinions *can* be changed. Opinions can be supported by facts: one person may form one opinion when presented with certain facts, but another person may interpret these facts differently and form a different opinion.

Analyse, interpret and evaluate the evidence

→ What evidence has the author included?

→ Where is the evidence from?

→ Could the evidence be interpreted in a different way?

→ Is there a bibliography? If so, look at the books or journals included in the bibliography and evaluate their quality.

Evaluate the conclusions

→ Are the conclusions valid? (Does the evidence lead to these conclusions?)

→ Could different conclusions be drawn from the evidence?

Annotated example

Drinking alcohol is good for you!

August 2017

A daily alcoholic drink could reduce your risk of dying young by 25%.

Experts have discovered that drinking a pint of beer or a glass of wine every day reduces your risk of heart disease.

A study has found that women who drink three to seven alcoholic drinks a week are 25% less likely to die prematurely. Their chance of dying from heart disease was 34% lower than lifelong abstainers. Men who consume three to 14 alcoholic drinks a week are 13% less likely to die prematurely and reduce their risk of dying from heart disease by 21%.

Drinking small amounts of alcohol is thought to improve blood vessel health by reducing inflammation, and to increase the levels of 'good' cholesterol in the blood.

However, the *Journal of the American College of Cardiology* reports that drinking alcohol becomes harmful beyond a certain point.

Dr Bo Xi led a team of scientists from the Shandong University School of Public Health in China, who analysed data from 333 247 people between 1997 and 2009. They found that men who drink more than 14 alcoholic drinks per week have a 25% increased risk of dying prematurely and a 67% increased chance of dying from cancer.

Annotations:

An attention-grabbing headline. Some people who are skim-reading a newspaper might just read this headline and accept it as fact.

What is meant by 'young'?

Refer to original research and check that this is what the experts concluded.

Need to check who the 'experts' are that the journalist refers to.

All of this evidence suggests drinking is better for health than abstaining. But why are people lifelong abstainers? Is there an underlying health reason for abstaining from alcohol?

Use of the word 'thought' is a bit vague. Is there evidence?

The journal that published the research is named. This adds credibility, but also gives a starting point for finding the original article.

Sample size is large.

Drinking alcohol does not seem quite so 'good for you' as the headline suggests!

The original article published in the *Journal of the American College of Cardiology* can be easily found. If you read the original article, would you come to the same conclusion as the journalist and report than 'alcohol is good for you'? Or suggest that people should have a 'daily alcoholic drink'?

Activity

Find a science- or health-based story in the news that seems to make sensational claims. Regular reports are made about diets that will cure specific diseases, fitness fads, super foods or alcohol intake. Identify the source of the research behind the article and find the original journal article. Critically read the article and analyse the evidence. Would you reach the same conclusions as the journalist? Could the evidence be interpreted differently?

Reading exam questions

When approaching an exam question, remember to use the reading skills that you have developed:

→ **Skim-read** through the whole question and identify the main topic.

→ **Intensively** read the introductory information and underline or highlight key words, or write notes to support your understanding.

→ **Scan** each question part and identify the command word (this tells you what to do when answering the question). Understanding the command word is essential for structuring your answer and including the relevant points.

> ✅ **Exam tip**
>
> Make a glossary with all of the command words that could be used in the exams and refer to it when you are answering past paper questions. This will help you to plan your answers and include the points that the examiner is looking for.

Worked example 2.1

A scientist investigated the relationship between body mass and the rate of oxygen uptake in four different mammals.

The scientist's results are shown in Figure 2.2.

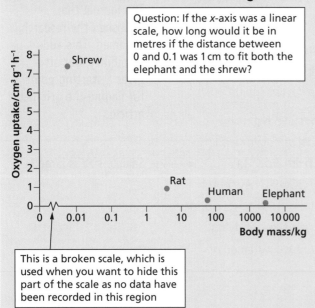

Question: If the *x*-axis was a linear scale, how long would it be in metres if the distance between 0 and 0.1 was 1 cm to fit both the elephant and the shrew?

This is a broken scale, which is used when you want to hide this part of the scale as no data have been recorded in this region

(a) The *x*-axis scale for body mass is a logarithmic scale. Give the reason for using a logarithmic scale. **(1)**

(b) Oxygen uptake is measured in $cm^3\,g^{-1}\,h^{-1}$. Explain why the scientist measured the rate of oxygen uptake per gram of body mass. **(1)**

(c) Describe the relationship between body mass and the rate of oxygen uptake. **(1)**

(d) Mammals maintain a constant body temperature. Use this information to suggest why the graph shows the relationship you described in (c). **(3)**

Figure 2.2 Graph to show the rate of oxygen uptake in mammals with different body mass

Step 1: Skim read through the question.

This question is about *oxygen uptake*. You should then make the link between oxygen uptake and *respiration*: *The higher the rate of oxygen uptake, the higher the rate of respiration.*

Step 2: Read the question intensively.

Include careful reading of the graph:
→ What scale has been used on the *x*-axis? Why is it a broken scale? Why does it increase in the increments shown?
→ What are the four mammals that have been investigated? Are there any similarities/differences between them?
→ What scale has been used on the *y*-axis? What do the units tell us?
→ Is there an overall pattern or trend?

Step 3: Scan for command words.

(a) 'Give' is the command word, so a simple statement is required in your answer:

A logarithmic scale allows you to fit the four different animals onto the same graph when their body masses vary considerably. (The shrew has a mass of less than 0.01 kg, whereas the elephant has a mass of approximately 3000 kg.)

(b) The command word is 'explain', so give a reason *why* the scientist measured oxygen uptake per gram. Answers to 'explain' questions cannot be found in the question — you need to use your scientific knowledge and understanding.

Measuring oxygen uptake per gram of body mass allows comparison between four animals of different mass.

(c) 'Describe' requires a simple statement linking the two variables and giving the overall pattern or trend shown by the graph:

As the body mass of the organism increases, the rate of oxygen uptake decreases.

(d) 'Suggest' means that you have to come up with a sensible scientific explanation for unfamiliar material. You have been given information about body temperature, so your answer needs to make a link between the size of the mammal, its rate of respiration and its body temperature.

See 'The difference between...' below.

The difference between...

A good student	An excellent student
(d) Smaller animals like the shrew have a high oxygen uptake of about 7.3 cm^3 g^{-1} h^{-1}, whereas large animals like the elephant have a much lower oxygen uptake of 0.2 cm^3 g^{-1} h^{-1}. The higher oxygen uptake of the shrew means that it must have a much higher rate of respiration than the elephant. Heat released during respiration is used to maintain the constant body temperature of the mammals. *The good student shows excellent understanding of the graph in Worked example 2.1 and makes good links between the information provided in the question and their own scientific knowledge. However, no mark is awarded for the first sentence as it is 'describing' the data and no mark is awarded for the last sentence as it doesn't suggest why the different mammals have different rates of respiration.*	(d) Smaller animals have a larger surface area to volume ratio, therefore, they have a higher rate of heat loss per gram of body tissue and require a higher rate of respiration to maintain a constant body temperature. The higher rate of oxygen uptake of smaller animals is because they are using more oxygen for their higher rate of respiration. *The excellent candidate realises that no marks are available for describing the data and so does not waste time or space on a description. All 3 marks are awarded for this answer as it makes clear yet concise links between oxygen uptake, respiration and heat loss.*

> ## ! Common pitfall
>
> Excellent students sometimes lose marks because they think that an answer seems too obvious, so they write something different. Do not overcomplicate exam questions. If the answer seems obvious, it is probably correct!

> ## ✓ Exam tip
>
> You have been provided with the information in a question for a reason. The information may be necessary for you to answer the question, or it may provide guidance for the answer that the examiner is looking for.

Worked example 2.2

Trypsin is a protease. Some of the cells of the pancreas produce an inactive form of trypsin. The inactive form of trypsin is secreted into the small intestine. In the small intestine, another enzyme removes a short chain of amino acids from the end of the inactive trypsin to produce the active form of trypsin.

Suggest the advantage of the pancreas cells producing an inactive form of trypsin.   (2)

Step 1: Skim read through the source material to identify the main topic.

This question is about enzymes and the digestive system.

Step 2: Read the stem of the question intensively.

Remember that actively reading means that you need to *do* something with the information. As you read, you could translate the information into a simple visual form that will help you to understand and process the content:

trypsin = protease

inactive trypsin (pancreas cells) → small intestine (chain of amino acids removed) → active trypsin

Step 3: Scan to identify the command word.

This is a 'suggest' question, meaning that the context may be unfamiliar to you, but you can use your scientific knowledge and understanding to write an explanation.

Although you may not have heard of trypsin, you are told that it is a protease.

The question asks you why it would be an *advantage* to pancreas cells to produce an *inactive* form of trypsin. What would happen inside the cell if trypsin was in an *active* form? Proteins inside the cell could be hydrolysed by this active form of trypsin, causing damage to the cell organelles.

The advantage to the pancreas cells of producing an inactive form of trypsin is that the trypsin does not hydrolyse protein inside the cell, which would cause damage to the cell.

Application to the exam

Sample question

This exam question assesses a number of the skills discussed in this chapter. There is guidance on how to approach the question, a mark scheme so you can self-assess your answer, and then a sample answer at the end.

A parasite is an organism that lives in or on another organism called the host organism. Parasites get their food from the host organism and often cause harm to the host.

Wing feather mites are parasites found in the wing feathers of many species of bird. The mites feed on an oil called preen oil produced by birds. Birds produce preen oil to maintain their feathers and keep them in good condition. Birds that cannot oil their feathers use more energy to maintain a constant body temperature, so less energy would be available for other processes.

Scientists investigated the relationship between the breeding success of one species of bird, greenfinches, and the number of feather mites counted in their wings.

Suggest how breeding greenfinches could be affected by feather mites. **(2)**

How to approach the question

This is a 'suggest' question with a lot of information to read through.

Remember to skim (identify the main topic), intensively read (identify the key points that will help you answer the question) and scan (identify the command word):

→ Birds need oil for healthy feathers.

→ Mites feed on oil.

→ Lack of oil → less available energy

These key points can be linked together to suggest how mites would affect breeding.

Mark scheme

1 reduced breeding success;

2 (more energy to maintain body temperature so) less energy for breeding;

Sample answer

This seems like an obvious thing to state, but the question asks how breeding is *affected*.

Feather mites lead to the feathers being in poor condition and make breeding *less successful.* ✓ Less energy is available for reproduction/mating/courtship displays. ✓

You could come up with other sensible suggestions of how less energy might reduce breeding.

Comprehension-style questions

Some exam questions require you to read a passage and then answer questions related to the passage. Although these are comprehension-style questions, you will not be able to find all of the answers in the passage. You are expected to use the information you are given, together with your own knowledge.

Read the following passage.

Some green plants are carnivorous. Carnivorous plants photosynthesise to provide a source of carbohydrate, but they also obtain some of their nutrients by feeding on animals.

5   **Carnivorous plants can survive in soil with a low nutrient content. Carnivorous plants are at a disadvantage compared with non-carnivorous plants in soil with a high nutrient content.**

Pitcher plants are carnivorous plants that capture insects in large extensions of their leaves called pitchers. Pitchers are deep and steep-sided with a solution in the bottom. Insects
10   **become trapped in the pitchers and cannot get out.**

Inside the pitchers are digestive glands. The cells of the digestive glands have a very similar ultrastructure to cells from the pancreas. The digestive gland cells firstly secrete enzymes to digest the trapped insects, then they absorb the soluble
15   **products of digestion. Scientists found that adding a respiratory inhibitor prevented the absorption of the products of digestion.**

Use information from the passage and your own knowledge to answer the questions.

(a) **Suggest why carnivorous plants are at a disadvantage compared with non-carnivorous plants in soil with a high nutrient content. (Lines 5–6)**   (1)

(b) **Cells of the digestive glands in pitcher plants have a similar structure to cells from the pancreas. Explain two adaptations of the digestive gland cells found inside the pitchers of pitcher plants. (Lines 11–13)**   (2)

(c) **Adding a respiratory inhibitor prevented the absorption of the products of digestion. (Lines 15–16)**

   Name the process that absorbs the products of digestion. Explain your choice using evidence from the passage.   (2)

How to approach the question

Skim reading through the passage tells you that the passage is about familiar content covered in the A-level specification:

→ plants

→ digestive glands

→ digestion

→ absorption

Intensive reading of the passage is then required to understand how the familiar A-level content relates to the unfamiliar idea of carnivorous plants, specifically pitcher plants.

Scan through each question to identify the command word.

(a) You need to **suggest** a reason *why* carnivorous plants are at a disadvantage.

In high-nutrient soil, the non-carnivorous plants can obtain nitrates from the soil. How would this affect their growth?

(b) You have to **explain** two adaptations, so you must state what each adaptation is *and* why it is necessary.

Digestive gland cells would have adaptations to produce large amounts of protein because they are secreting enzymes. They would also have adaptations to rapidly absorb products of digestion.

(c) You have to **name** a process and **explain** your choice.

You should already know that the products of digestion can be absorbed by diffusion — a passive process — or by active transport. The question tells you that adding a respiratory inhibitor prevents absorption, so you need to link this information to a named process of absorption.

> ✓ **Exam tip**
>
> Remember that respiration releases energy or produces ATP. Never state that energy is produced as energy cannot be created (or destroyed). You may be penalised in an exam for writing 'energy is produced'.

Mark scheme

(a) Answer either suggests advantage of high-nutrient soil to non-carnivorous plants or disadvantage to carnivorous plants:
  1 non-carnivorous plants grow faster so outcompete carnivorous plants; *or*
  2 carnivorous plants use more energy for enzyme synthesis/have a lower leaf area for photosynthesis so less energy available for growth;

(b) two adaptations stated, with explanation for each adaptation linked to protein synthesis;

(c) 1 active transport named;
  2 link made between use of ATP in active transport, ATP production by respiration and inhibition of respiration;

Sample answer

Marking point 1 would be awarded for explaining the advantage to non-carnivorous plants.

The student could instead have explained the disadvantage to carnivorous plants for marking point 2. For example, carnivorous plants require energy to synthesise enzymes and absorb nutrients, so less energy is available for growth.

(a) In high-nutrient soil, non-carnivorous plants can obtain the nitrogen that they need in the form of nitrates from the soil. They would grow quickly and outcompete carnivorous plants. ✓

(b) The cells would have high numbers of mitochondria to produce the ATP needed for enzyme synthesis. ✓ There would also be high numbers of ribosomes and rough endoplasmic reticulum (ER) to synthesise the enzymes needed for digestion. ✓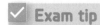

The process is named and linked to clear evidence from the text, so both marks are awarded.

(c) Active transport requires ATP produced during respiration. The text states that absorption is prevented by a respiratory inhibitor, so the process must be active transport ✓ as it requires ATP from respiration. ✓

Two adaptations have been given and linked specifically to the role of the cells in protein synthesis. The student has stated that the numbers of mitochondria and ribosomes are *high*. It would not be an adaptation to just state that mitochondria and ribosomes are present without referring to the high numbers of them. Other adaptations could be:

→ extensive endoplasmic reticulum for transport of enzymes
→ extensive Golgi for modification and packaging of the enzymes

Using your further reading in the exam

Having read extensively around the subject, you will be keen to show off your further reading in the exams. The best time to demonstrate your wider knowledge is when you are answering extended writing or essay questions.

AQA has a requirement for further reading that means that you can only access the top marks for your essay if you have included evidence of reading beyond the specification. The most important thing to remember is only to include content if it is *relevant*; do not include random information that is unrelated to the title as you may be penalised for including irrelevant content.

Common pitfall

Further reading is an excellent way of increasing your breadth of knowledge, but do not become too focused on trying to include evidence of this reading in the exams. Your answers to exam questions must reflect the specification content.

You should know

> Remember that your core textbook is a useful resource for independent study. Make the most of the extension activities and summary questions in it.

> When you read an exam question, identify the command word and *use* the information you are given in the question.

> Do not believe everything you read. Be critical and think more deeply about the source, evidence and conclusions.

> Enjoy your further reading. There are lots of books, journals and websites available, so you will definitely be able to find something that inspires you.

3 Writing skills

Your entire A-level grade is dependent on your ability to express yourself through your written work. It does not matter if you understand all of the content and can answer every question that your teacher asks you, if you do not write a clear answer in the exam, you will not get the mark. The examiner does not 'know what you mean' — you have to write it.

This chapter addresses the skills needed to succeed in the exams, with a focus on the different styles of exam question, including short-answer, extended-response and essay questions. For each style of question there is guidance on how to structure your responses and worked examples.

Study skills

Understanding command words

Many exam questions are short-answer questions worth 1–3 marks. Some of these questions require very brief responses of one word or phrase, with no requirement for a full sentence. This type of question usually starts with one of these command words:

→ Give
→ Identify
→ Label
→ Name
→ State

For example, you could be told that a student is investigating the effect of temperature on the rate of an enzyme-controlled reaction and asked to '*Give two variables that the student should control*'.

The command word here is 'give' and a suitable response would be:

→ pH
→ substrate concentration

There is no need to 'explain' *why* the variables were controlled or to 'describe' *how* the student controlled the variables, just *give* the variables.

 Exam tip

Exam questions may ask for a specific number of examples, for example, name *two* products of lipid hydrolysis. Check that you have included the correct number of examples; there is no need to add extra ones.

 Exam tip

Beware of the 'list rule' in exams: if you write two correct answers followed by an extra incorrect one, the incorrect answer will cancel out one of the correct answers. For example: 'Name *two* factors that affect the rate of photosynthesis':

→ temperature ✓
→ light intensity ✓
→ oxygen concentration ✗
Only 1 mark out of 2 is awarded despite the first two answers being correct.

Some exam questions require you to write in sentences, including those that start with the command words:

→ Compare
→ Contrast
→ Define
→ Describe
→ Evaluate
→ Explain
→ Suggest

These questions require you to demonstrate your knowledge and understanding of the subject. You need to write your answer using the correct technical language so that it sounds like you know what you are talking about. Biology requires the use of very specific terminology. An A-grade student will use the correct scientific terms as part of a fluent and concise answer.

✓ Exam tip

Check that your answer does not repeat information given in the stem of the question because there will be no mark available for this. For example: 'Humans have millions of alveoli to provide a large surface area for rapid gas exchange. Give two other features that cause rapid gas exchange in humans.' There is no mark available for repeating 'millions of alveoli' or 'large surface area'.

Table 3.1 Some commonly used command words and their meanings

Command word	Meaning
Compare	Identify similarities and/or differences. You may need to include comparative terms such as: faster/slower; higher/lower; more/less.
Contrast	Identify the differences between the items.
Define	Give the meaning of the term. Try to learn definitions as they are given in the specification.
Describe	Describing data requires you to give the overall pattern or trend and use the data. For example: 'the rate increases steadily and then plateaus at $3.2\,cm^3\,min^{-1}$'. Describing a method or process means giving the main steps or features.
Evaluate	Make a judgement using the available evidence. You may need to consider pros and cons.
Explain	Use your scientific understanding to give reasons for results or biological phenomena.
Justify	Use the information given and support a case with reference to the evidence.
Suggest	Give a possible cause or a reasonable explanation of the information. There may be more than one correct response.

Worked example 3.1

You may be asked to describe experimental techniques that you have followed in class. This type of response must be sequenced logically. Try to imagine yourself completing the practical activity, then write a set of instructions that another student could use to replicate the practical. Use imperative ('bossy') verbs at the start of your sentences to give instructions, for example, measure, heat, add.

Describe how you could show that a sample of cheese contains lipid. (2)

Crush the cheese and shake it with ethanol, then add water. ✓ *A white emulsion* ✓ *will appear if lipid is present.*

This question is asking you to describe a procedure that you may have completed during a practical session. The order in which these steps are completed is important because you will not obtain the correct result if water is added first.

The difference between...

Students frequently lose marks because of poor use of terminology when answering exam questions about enzymes, for example when writing about the effect of a non-competitive inhibitor:

A-grade response	C-grade response
The non-competitive inhibitor binds to an allosteric binding site away from the active site, causing the active site to change shape. The active site no longer has a complementary shape to the substrate, so the substrate cannot bind and an enzyme–substrate complex does not form. *The A-grade answer is clearly written with excellent use of terminology, such as 'allosteric', 'complementary' and 'enzyme–substrate complex'.*	A non-competitive inhibitor binds away from the active site. This changes the active site, so the substrate no longer fits. *The C-grade answer shows clear understanding of non-competitive inhibitors and their effect on enzymes, but lacks the correct scientific terms.*

An important consideration when answering an exam question is how much to write. You will be given a lot of space to answer your question because the exam board wants your answer to fit into the box. This does not mean that you have to fill all of the available space. It is important to know when to *stop* writing.

A 2-mark question should not require more than two sentences. In some cases, one perfect sentence could be worth 2 or 3 marks. For example, if you were answering a question about the human breathing system and were asked to describe how oxygen from the air reaches a red blood cell in the capillaries surrounding the alveoli:

Oxygen in the air moves down a pressure gradient ✓ *through the trachea, bronchi and bronchioles* ✓ *to the alveoli, then crosses the alveolar epithelium* ✓ *and capillary endothelium* ✓ *to reach the red blood cell.*

This one sentence could be worth up to 4 marks.

Quality of written communication

A high standard of written communication is expected from A-level students. Some sections of the exam paper may specifically award a mark for your quality of written communication (QWC).

You are expected to use scientific terminology correctly, and you should learn the spellings and definitions of key terms as part of your revision because you will not have access to spell check in the exam! Phonetic misspellings of many key words would be accepted — for example, students often misspell protein as 'protien', but it is obvious what is meant. Certain terms must be spelt correctly, however, because they could be confused with similar words. For example, 'meitosis' would not be accepted because it is an amalgamation of meiosis and mitosis.

Correctly used punctuation helps to structure your ideas and makes it easier for the examiner to understand your point. A poor standard of written work detracts from the content, so decide what you are going to write before you start writing. This is especially important when writing the essay — a full description of how to plan your essay is given later in this chapter. The examiners want to give you marks, so write your answer clearly and make the examiners' job easier.

 Exam tip

As a general rule, you have between 1 and 1.5 minutes per mark in the exam, so keep an eye on the time. If a question is taking too long, write a brief answer, then put an asterisk next to the question and come back to it at the end if you have time.

Planning and structuring answers

There are many different types of exam question so knowing how to answer each type of question is the key to exam success. Some questions require calculations and mathematical skills; this type of question is discussed in chapter 1 (Quantitative skills).

The main types of question requiring writing skills include:

→ short-answer structured questions
→ extended-response questions
→ essay questions

The types of question and suggested methods for approaching them are described in this section.

Short-answer structured questions

Short-answer questions generally account for the majority of the marks in a written exam, and may assess:

→ factual recall
→ knowledge and understanding of the specification
→ ability to apply knowledge

Your understanding of command words is essential when determining what you need to write and how much to write.

Little planning is required for this type of question because your answer will be 2–3 sentences at most.

Worked example 3.2

Explain how tissue fluid is formed. (2)

Once you have identified that an explanation is required to answer this question, make sure that you do not add additional information beyond the *formation* of tissue fluid. No marks will be available for describing the return of tissue fluid to the capillary.

> The hydrostatic pressure at the artery end of the capillary is high ✓ and this forces small molecules out of the capillary ✓ to form tissue fluid.

Some students would focus on the phrase 'tissue fluid' and then write down everything that they know about tissue fluid.

Worked example 3.3

Explain how high blood pressure can lead to an accumulation of tissue fluid. (3)

This example is more demanding than worked example 3.2 because it is asking you to *apply* your knowledge to a novel situation — a person with *high* blood pressure. Throughout your answer, you need to focus on what would be *different* about tissue fluid formation in a person with high blood pressure that would lead to tissue fluid accumulating. Your answer needs to include comparative terms such as *more*, *higher* and *increased*.

> High blood pressure causes a *higher* hydrostatic pressure ✓ in the capillary. This leads to an *increased* outward pressure ✓ from the capillary at the artery end, so *more* tissue fluid is formed. Tissue fluid is formed *faster* than it can be reabsorbed ✓, leading to an accumulation.

> **!** **Common pitfall**
>
> Messy and illegible writing can lose you marks in an exam. If the examiner cannot read what you have written, then you will not be awarded the mark. Use a good quality black pen and write as clearly and neatly as you can. If you make a mistake, cross it out neatly and rewrite your answer.

Extended-response questions

Extended-response questions may assess core skills of knowledge and recall. These questions will have command words like 'describe' and 'explain'. When you are answering a question like this, decide on a logical order and then write in short sentences (not bullet points).

Plan and structure your answer to enable you to achieve full marks. Four sentences should be sufficient for a 4-mark question.

Worked example 3.4

Describe the digestion of starch. (4)

Step 1: Describe the digestion of starch in the mouth using scientific terminology — 'hydrolyse' rather than 'break down'; named bonds rather than just 'bonds'.

Step 2: Describe starch digestion in the small intestine.

Step 3: Describe maltose digestion in the small intestine.

Starch digestion begins in the mouth with chewing (to increase surface area) and the addition of saliva to food. Saliva contains salivary amylase ✓, an enzyme that hydrolyses the glycosidic bonds in starch and produces maltose (a disaccharide). ✓

Salivary amylase is denatured in the stomach and then starch digestion continues in the small intestine where pancreatic amylase hydrolyses the remaining starch to maltose. ✓

Maltase, bound to the membranes of the intestinal epithelium, hydrolyses maltose to α-glucose. ✓

Once maltose has been hydrolysed to α-glucose, starch digestion is complete. There is no need to describe absorption.

Worked example 3.5

Describe how light energy is used during the light-dependent reactions of photosynthesis. (5)

Step 1: Describe how light energy affects chlorophyll and how the energy is used.

Your answer must focus on the role of light energy rather than writing generally about the light-dependent reactions.

➥

Step 2: Describe how light energy affects water.

A chlorophyll molecule absorbs light energy and this energy excites electrons. ✓ The electrons are taken up by an electron carrier and pass along an electron transfer chain, releasing energy ✓ as they pass from carrier to carrier. The energy released is used to combine a molecule of ADP with inorganic phosphate to produce ATP. ✓

Light energy is also used in the photolysis of water in which a molecule of water is split to produce protons, electrons and oxygen. ✓ The protons from the photolysis of water and the electrons from the chlorophyll molecule are taken up by an electron carrier called NADP to produce reduced NADP. ✓

There is no need to go into detail about the chemiosmotic theory in your answer, because the question just asks you to describe *how light energy is used* and not how ATP is produced.

In summary, when answering extended-response exam questions:

1 Focus on the *command word* — what does the command word tell you to do?
2 *Plan* what you need to say to answer the question, and sequence your ideas logically.
3 Write a *clear and concise* answer that avoids repetition or unnecessary detail.

> ✓ **Exam tip**
>
> In an exam inorganic phosphate can be written as 'P_i', which is used to mean 'inorganic phosphate', but not just as 'P', because a molecule of ADP combines with phosph*ate*, not phosph*orus*.

Essay questions

Some exam papers include an essay question worth up to 25 marks. You should allow yourself 35–40 minutes to complete a 25-mark essay and, although essays are often at the end of the paper, you do not have to do them last. As long as you are strict with your timing and do not spend the entire exam writing your essay, you may prefer to complete the essay first.

Students are often overly concerned about the essay, but it is one of the most straightforward questions on the A-level paper. Essays assess the AO1 skill of demonstrating your knowledge and understanding, and the AO2 skill of applying the relevant subject knowledge to answer the question, but there is no AO3 assessment of applying, interpreting or evaluating. Although the essay question should be undemanding for the A-grade student, an 'exceptional' essay is difficult to write. Writing an exceptional essay requires you to make synoptic links, and it is *creating* these links that is the higher-order thinking skill.

The exam boards' marking criteria for essays tends to be more open-ended than they are for structured questions, and generally use a 'level of response' style of marking. This provides students with the opportunity to demonstrate the depth of their understanding, and is a way of identifying the more able students.

The difference between...

B-grade essay	A-grade essay
The answer will have clearly explained and linked topics that relate to the main theme of the question. The content will be detailed and of the standard expected at A-level, with the correct use of scientific terminology.	*The answer will include inter-linked and clearly explained topics that relate clearly to the theme of the question. The answer will be very well written with the correct use of scientific terminology and no factual errors or irrelevant content. For AQA, the answer will include evidence of reading beyond the specification to achieve top marks in the highest mark band.*

The best way to prepare for the essay is to identify the main themes throughout the specification and then create spider diagrams or mind maps linking topics and ideas together.

The following list suggests some of the main themes that may link together topics in the specification. It is not an exhaustive list — you may find other themes, and you will also find overlaps between some of these themes. For example, proteins will also be referred to in membranes and transport.

→ Disease
→ Energy
→ Exchange
→ Membranes
→ Proteins
→ Response
→ Transport

> **! Common pitfall**
>
> A-grade students will be well prepared for the exam and, as well as answering the essay questions from past papers, they may also have devised their own essay titles. Take care in the exam to answer the essay title you have been *given* rather than the essay title that you *want* or have practised for.

Worked example 3.6

You may be given a choice of essay titles in the exam, so make sure that you are familiar with the structure of the exam papers:
→ Which paper includes the essay?
→ How many essays do you have to write?
→ How many marks is the essay worth?
→ How much time should you spend writing the essay?

Step 1: Select the essay title.

The first step is choosing the title. Read through the choice of titles carefully and decide which one you are going to select. This choice may be obvious and you will know straight away which title you know the most about, or you may have to think more carefully about your choice.

Step 2: Analyse the essay title.

When *analysing* the title, you need to decide what you need to include:
→ Is there a command word in the title of the essay?
→ Is the title a question that requires you to discuss evidence and reach a conclusion?
→ Are you asked to write about a specific organism, or organisms in general?

> ✓ **Exam tip**
>
> An essay about 'living things' or 'organisms' requires examples from the full range of organisms that you have studied, and should not just be about humans.

Activity

Look through a past paper and choose an essay title. Analyse the title and identify relevant topics to include, then create an essay plan following the guidance given in this chapter. Now look at the mark scheme for the essay (this may be a list of possible topics rather than a comprehensive answer) and check your plan — have you included a suitable number of topics? Repeat this planning process with other essay titles from past papers.

Step 3: Plan the essay.

It is tempting to rush straight into writing the essay, especially as you are under exam conditions and have a time constraint, but better students draw up a plan before they start writing. A plan helps to keep the content relevant, ensures the requisite breadth and focuses your writing — you will not be able to write everything that you know and will need to focus on the most relevant examples. If you only have about 30 minutes to write the essay, realistically you can only write five or six paragraphs.

When you first look at the title, lots of thoughts about what to include may spring to mind — jot down these random thoughts and ideas as you have them. For example:

Write an essay about the importance of cycles in biology.

Think about all of the cycles you have studied throughout the course, jot them down and see if you can make links between them. Your specification may have included:
→ ecological cycles — carbon, nitrogen, phosphorus and water
→ biochemical cycles — Krebs, Calvin and ornithine
→ physiological cycles — menstrual, cardiac
→ the cell cycle
→ the polymerase chain reaction (as a three-step cycle)
→ lytic cycles (of viruses)

Now identify the main topics that you are going to write about by selecting the topics that you know most thoroughly and feel most confident writing about. There is no need to write about all four ecological cycles, so just choose one to describe in detail and the others can just be mentioned. (Just writing about ecological cycles would produce an unbalanced essay that did not have the required breadth of coverage of the specification.) Similarly, you do not need to write about all three biochemical cycles — just one or two.

Decide on a logical way of sequencing the topics, and then number them. The aim of this sequencing is so that the essay will 'flow'. There is no set way to determine this order — just whatever seems logical to you. One example of an order is given and explained below:

1 Calvin — cycles begin at a *sub-cellular* level; the Calvin cycle occurs within the chloroplasts of plant and algal cells.
2 Cell — moving up to a *cellular* level, the cell cycle is the regular cycle of cell growth and division of a cell.
3 Lytic — also at a *cellular* level, the lytic cycle describes the destruction of a cell infected with viruses.

> **! Common pitfall**
>
> Excellent students sometimes try to include too many examples or too much information in their essays and end up writing content that is too superficial. The content you include must be written with the level of detail expected at A-level.

> **✓ Exam tip**
>
> Appropriately used diagrams can be very useful when answering exam questions. Including a *clearly drawn*, *labelled* and *relevant* diagram can be easier than trying to describe something — for example, the fluid-mosaic structure of a membrane, or countercurrent flow.

4 Cardiac — at an *organ* level, the cardiac cycle is the sequence of contraction and relaxation of the atria and ventricles.

5 Nitrogen — at an *ecosystem* level, the nitrogen cycle describes the flow of nitrogen between the atmosphere and living things.

The order described works its way up in terms of size from sub-cellular to ecosystem level. If the order makes sense to you, then you will be able to make links between the paragraphs to give your essay the necessary 'flow'.

Before you start writing the essay, look at the title again and check your plan:
→ Is everything in your plan *relevant* to the title?
→ Have you answered the question (if it was a question)?
→ Is there a *breadth* of examples from across different sections of the specification?
→ Have you identified an opportunity to include evidence of your *further reading* if necessary?

Planning your essay for 5 minutes is time well spent because it will help you to write a well-structured essay with good flow and a logical sequence of ideas.

Step 4: Write the essay.

Start your essay with a brief introduction. This only needs to be a few lines introducing the main ideas you will be discussing in your essay and indicating how these ideas link to the title.

The main body of your essay will consist of five or six paragraphs about the topics you identified in your plan. These paragraphs must be explained in detail and you must stick to your plan to avoid drifting into irrelevant content. For example, if you are writing about the Krebs cycle, it is relevant to *mention* glycolysis and the link reaction as providing the raw material for the Krebs cycle, and to *mention* the electron transfer chain when referring to the products of the Krebs cycle, but describing these processes in detail is irrelevant to the title.

When moving from one paragraph to the next, you need to make clear links so that the paragraphs are not completely isolated ideas. For example, if you had just written about the Krebs cycle and were starting a paragraph on the Calvin cycle, you could write:

> The Calvin cycle in the stroma of chloroplasts is another example of a biochemical cycle, but whereas the Krebs cycle *produces* ATP and CO_2, the Calvin cycle *uses* ATP and CO_2.

Finish your essay with a brief conclusion that summarises common features or ideas that you have discussed in your essay. For example:

> Throughout this essay I have described biological cycles, from the sub-cellular biochemical processes of respiration and photosynthesis to vast ecological cycles such as the nitrogen cycle. I have shown that the key principle of cycles in biology is their continuous nature; there is no specific start or end point, because the end point of one process provides the start of the next.

Step 5: Review your essay.

Leave yourself about 5 minutes to read through your essay. As you review your essay, remember to consider:
→ QWC — is your essay well written, with excellent spelling, punctuation and grammar? Have you used scientific terminology correctly? Are there any factual errors?

✓ Exam tip

Presentation is important when writing your essay. Space out your work to make it easier to read, and leave a line between each paragraph in case you want to add extra information after you have reviewed your essay.

✓ Exam tip

Including an introduction and conclusion is not an exam board requirement, but is a good essay writing technique.

✓ Exam tip

Remember that your quality of written communication must be of a high standard throughout the essay. Poor spelling, punctuation and grammar will detract from the content of the essay.

→ relevance — is all of the content directly related to the title?
→ synopticity — have you included examples from across the specification, with clear links made between topics?
→ extension — have you included evidence of your further study/reading for AQA? (This material *must* be A-level standard.)

Activity

Choose an essay title from a past paper and then revise all of the relevant content from your specification. Refer to a magazine like *Biological Sciences Review* to find and read extension material that relates to the essay title. Then find a quiet place to work where you will not be disturbed and hand-write the essay under timed conditions. Either self-assess your essay using the mark scheme, or ask a friend or teacher to assess your work. (This would be preferable as you may be too lenient if you are marking your own work!) Make this timed essay practice a regular feature of your independent learning.

Take it further

Your knowledge outside of the specification does not just have to come from your reading. There are some exceptional documentaries that give fantastic insights into a wide range of living organisms and their unusual adaptations. David Attenborough has written, narrated and presented a huge number of outstanding documentaries exploring life on this planet.

Application to the exam

Extended-response questions

Antibiotics are added to animal feeds to prevent the spread of disease between intensively farmed animals. The UK is trying to reduce the use of antibiotics in animal feeds. Suggest reasons why. **(4)**

How to approach the question

Your A-level specification may have included content on the development of antibiotic resistance in bacteria by natural selection, and you may also have covered intensive farming methods. This synoptic question is asking you to use this knowledge from across the specification to come up with sensible suggestions. Look at the number of marks available to help you decide how much to write. As there are 4 marks available, aim for four suggestions.

Give yourself 6 minutes to answer this question.

! Common pitfall

Do not think that you have to fill all of the available space allocated to a question. Students sometimes end up writing too much, and end up losing marks by contradicting themselves or making mistakes.

Mark scheme

1 increased numbers of antibiotic-resistant bacteria;
2 antibiotic-resistance gene could be transferred to pathogens;
3 treatment for antibiotic-resistant pathogens less effective;
4 (antibiotics may cause) side-effects in animals;
5 (antibiotics) present in food for human consumption;
6 increased public demand for organic food;
7 antibiotics are expensive;
8 antibiotic use controlled by EU/government legislation;

> ✓ **Exam tip**
>
> In the margins of an exam paper, you may notice the phrase 'Do not write outside the box'. Do not write outside the box! Most exam papers are scanned and marked by examiners using online software. If you write outside the space allocated to answer the question, your answer may not be seen by the examiner.

Sample answer

Correct use of scientific terminology.

Widespread use of antibiotics can lead to an increase in the number of bacteria developing antibiotic resistance. ✓ The gene for antibiotic resistance can be passed on by horizontal transmission to other bacteria, including pathogens ✓ that cause human disease. This can result in the development of 'superbugs' which cause diseases that cannot be treated with antibiotics. ✓ For these reasons, government guidelines have been introduced ✓ to try to reduce antibiotic use.

There is no need to go into detail about *how* antibiotic resistance develops; just focus on reasons why it is a good idea to reduce antibiotic use in animals. Remember that resistance is the result of a random mutation and is not caused by the antibiotics themselves.

Essay question

Write an essay on the importance of membranes in different types of cell. (25)

How to approach the question

This essay title is about the *importance* of membranes, so the essay must focus on the roles of membranes and the functions that are linked to them. The title also requires reference to different *types* of cell, so consider the cells that you have studied in different parts of the specification. The essay requires you to make synoptic links between different topics, not just write in a lot of detail about one or two topics.

There is much that you could write about if given this title, so you should see the essay as an opportunity to cherry-pick the areas of the specification that you are most confident with and to showcase your knowledge. There is a lot of content that would be relevant to this title, but you cannot possibly write about every topic, and to do so would disadvantage you because the level of detail would not reflect the depth studied at A-level. Within a topic, you cannot include everything that you have studied, only the parts that are directly related to the theme. The main thing to remember is that all material that you introduce must be clearly linked to the title to show how it is relevant.

You need to write a brief introduction, followed by five or six paragraphs about linked topics, and finally a brief conclusion. Choose your five or six topics from different areas of the specification, then just write about the specific section of that topic that relates to the title. For example, including the thylakoid membranes of chloroplasts would be relevant as long as you described the location of photosynthetic pigments and electron transfer chain proteins within the membrane. It would *not* be relevant to start describing any of the light-independent reactions in the stroma, as these are not occurring on membranes.

Give yourself 35 minutes to plan and write the essay.

Once you start writing the essay, keep referring back to your plan so that you do not include irrelevant content. Keep an eye on the time — you only have about 5 minutes to write each paragraph.

Mark scheme

Mark schemes for essays use level-of-response marking. This means that the examiner will identify the mark band that describes your essay and then allocate a mark within that mark band. These levels of response vary slightly between exam boards, so check the mark schemes to see how your exam board assesses these essay questions. The list below includes some suggested content, but examiners may accept other relevant examples.

1   functions of membranes — selective permeability, transport across membranes;

2   membranes in organelles — thylakoids of chloroplasts, cristae of mitochondria, ER, Golgi;

3   cell surface membranes — receptors, antigens and immune response, Pacinian corpuscle;

4   neurones and action potentials — channel proteins, sodium-potassium pump;

5   synaptic transmission — vesicles, receptors;

6   muscle contraction — neuromuscular junction, calcium ion channels;

7   hormones — second messenger model, adrenaline and glucagon;

Sample plan

1   Introduction
   → Where membranes are found — around cell, around organelles, inside organelles
   → Functions of membranes in these locations
2   Membrane structure
   → Phospholipid
   → Proteins — extrinsic and intrinsic
   → Fluid-mosaic
3   Cell-surface membrane in all cells
   → General function — selectively permeable
   → Diffusion across lipid bilayer
   → Facilitated diffusion through channel and carrier proteins
   → Active transport using carrier proteins (refer to co-transport if time)

4 Neurone — resting and action potentials
  → Membrane impermeable to Na⁺ and K⁺
  → Na⁺/K⁺ pump
  → Na⁺ and K⁺ ion channels
5 Pacinian corpuscle
  → Stretch-mediated Na⁺ channels
  → Pressure opens channels
6 Membranes inside organelles
  → Mitochondrial membranes and cristae
  → Electron transfer chain
  → Creation of proton gradient
  → Proton flow and ATP synthase
7 Hormones and receptors on target cells
  → Adrenaline and liver cells
  → Transmembrane protein with adenyl cyclase
  → Second messenger model
8 Conclusion

Sample answer

Membranes are essential structures found in and around all cells, whether eukaryotic or prokaryotic. They are found around the cell as the cell surface membrane, separating the cell contents from the external environment. They are also found around organelles such as mitochondria and lysosomes, where they allow separate compartments to exist within a cell. In lysosomes, this compartmentalisation means that hydrolytic enzymes that could damage the cell are unable to cross the membrane and are kept isolated. Membranes within organelles form surfaces on which reactions can take place, like the cristae in mitochondria, where some of the reactions of respiration occur on the membrane. This essay will discuss the structure of these membranes and how important this structure is for the membranes to function. ←

The introductory paragraph identifies the main theme of the question (the *importance* of membranes) and suggests how the essay content is going to link to this theme.

The start of this paragraph flows directly from the content in the introductory paragraph to show how it is relevant to the title.

→ All membranes have the same general structure, whether the membrane is around a cell, around an organelle or within an organelle: they all have a phospholipid bilayer structure, with embedded proteins. The phospholipid bilayer allows lipid-soluble material to pass through it so that substances can enter and leave, while preventing water-soluble substances from passing through. The phospholipids can move, giving the membrane a flexible structure. Proteins are distributed throughout the phospholipid bilayer and can be found as extrinsic proteins on the inner or outer surface of the membrane, and as intrinsic proteins that completely span the membrane. Proteins found at the outer surface may have roles as receptors, or in adhesion between adjacent cells. Proteins spanning the membrane are called transmembrane proteins and may be channel proteins (fluid-filled channels that enable water-soluble molecules to enter or leave the cell by diffusion) or carrier proteins (proteins that bind to ions or molecules and then change shape to transport substances across the membrane). ◄——

> This paragraph explains membrane structure in detail, with excellent use of scientific terminology and a very clear application to the title.

One of the key functions of a membrane is selective permeability, which gives control over the movement of substances into and out of a cell or organelle. To exert this control, an important property of the cell-surface membrane is that it is not readily permeable to molecules. Small, non-polar molecules, for example oxygen, can diffuse through the phospholipid bilayer, but polar molecules and charged ions cannot diffuse through membranes dues to the hydrophobic nature of the hydrocarbon tails in the phospholipids making up the bilayer. These polar or charged molecules can enter and leave cells via transmembrane proteins through a process called facilitated diffusion. Some protein channels are selective to a specific ion and only open if that ion is present, allowing the ion to pass through a fluid-filled hydrophilic channel. The presence of these protein channels in membranes is important in the control of ion movement and the maintenance of electrochemical gradients. Molecules can also enter and leave cells directly or indirectly by active transport, which transports an ion or molecule against its concentration gradient using energy in the form of ATP. Active transport uses proteins in membranes to act as carriers, for example, the sodium-potassium pump. ◄——

> This sentence makes the link to resting potential in the following paragraph.

> Throughout this paragraph, the importance of the phospholipid bilayer and the membrane proteins in the movement of molecules has been emphasised to apply the content to the title. The reference to electrochemical gradients and the sodium-potassium pump leads on nicely to the next paragraph on neurones.

Neurones are specialised cells found in the nervous system; their function is the rapid transmission of nerve impulses from one part of the body to another. Nerve impulses are waves of electrical activity that travel along the axon membrane. The phospholipid bilayer of the axon membrane prevents the charged ions Na^+ and K^+ from diffusing across it. Embedded within the axon membrane are specific protein channels that allow either Na^+ or K^+ to move through by facilitated diffusion. There are also carrier proteins in the axon membrane that actively transport three Na^+ out of the axon for every two K^+ they transport into the axon. The impermeability of the lipid bilayer to Na^+ and K^+, together with the active transport of more Na^+ out than K^+ in, maintains the resting potential. In a resting neurone, the inside of the axon is negatively charged relative to the outside of the axon. When a receptor, for example a Pacinian corpuscle, detects a stimulus, the energy of the stimulus causes voltage-gated

> Using the term 'impulses' is an example of good use of scientific terminology, whereas 'signals' or 'messages' would be inappropriate terms to use.

> Naming the Pacinian corpuscle makes a good link to the next paragraph.

This topic is from the same section of the specification as the topic in the previous paragraph, but is a good example of a specific named cell, and is clearly applied to the title.

Na$^+$ channels in the membrane to open. Sodium ions rapidly diffuse into the axon, triggering more Na$^+$ channels to open and resulting in a reversal of the charges on either side of the axon membrane, so that the inside now has a positive charge. This is known as an action potential and this part of the membrane is now depolarised.

A Pacinian corpuscle is a sensory receptor found deep in the skin that responds to mechanical stimuli such as pressure. The membrane of a Pacinian corpuscle contains stretch-mediated sodium channels. In the normal resting state, these channels are too narrow to allow Na$^+$ to diffuse through them, which maintains the resting potential of the Pacinian corpuscle. Applying pressure to a Pacinian corpuscle stretches the membrane and increases the diameter of the sodium channels. Sodium ions are then able to diffuse across the membrane, causing depolarisation and creating an action potential.

This paragraph shows excellent understanding of the role of the membrane in the maintenance of the resting potential and the creation of an action potential. The level of scientific content and the use of terminology is of a high standard throughout. The student could have written a lot more about action potentials, but writing too much about one topic would limit the opportunities to make synoptic links.

Maintaining the resting potential across a plasma membrane is dependent on the sodium-potassium pump — the carrier protein that actively transports Na$^+$ out and K$^+$ in. Active transport requires ATP to change the shape of the carrier protein and allow movement across the membrane. ATP is synthesised in both animal and plant cells by oxidative phosphorylation in mitochondria. Mitochondria are surrounded by a double membrane and the inner membrane is highly folded to form cristae. The cristae provide a large surface area for the attachment of proteins and enzymes involved in oxidative phosphorylation. Prokaryotes, which lack membrane-bound organelles, have structures formed from their cell-surface membrane called mesosomes, where oxidative phosphorylation occurs. Oxidative phosphorylation involves the transfer of electrons down a series of electron carriers called the electron transfer chain. These electron carriers, including coenzymes and cytochromes, are embedded in their specific order within the inner mitochondrial membrane. Passing the electrons from carrier to carrier releases energy that is used to actively transport protons across the inner mitochondrial membrane and into the intermembrane space. This active transport of protons is dependent on a carrier protein within the inner mitochondrial membrane. The impermeable phospholipid bilayer prevents diffusion of the charged hydrogen ions back into the matrix and creates a proton gradient across the inner mitochondrial membrane. Protons move back into the matrix through protein channels that span the inner mitochondrial membrane. The flow of protons through

This is an excellent start to the paragraph because it explains how the previous material about the sodium-potassium pump relates to the mitochondria.

This reference to mesosomes reflects content that may not be on the A-level specification.

This is a good link made to the role of the membrane in maintaining the positions of electron transfer chain proteins.

Prokaryotes are examples of other types of cell, as required by the title.

This is evidence of high-level knowledge, beyond the limitations of the specification.

A very good link to the importance of membranes in maintaining a proton gradient.

these channels changes the structure of the enzyme ATP synthase and catalyses the synthesis of ATP.

This entire paragraph has focused on oxidative phosphorylation, because this is the only stage of respiration that requires a membrane. Avoiding references to glycolysis and the Krebs cycle has kept the content relevant to the title.

The transmission of nerve impulses referred to earlier in this essay is one form of coordination in animals, but animals also produce chemicals called hormones to coordinate responses. Hormones are transported in the blood and bind to their target cells via specific receptors on the cell-surface membrane, which have a complementary shape to the hormone. The receptor site is part of a transmembrane protein spanning the cell-surface membrane of the target cell. Adrenaline is a hormone that binds to receptors on the outside of liver cells. This binding causes the shape of the protein on the inside of the membrane, an enzyme called adenyl cyclase, to change into its active form. Activated adenyl cyclase can then initiate a series of reactions that lead to the conversion of glycogen to glucose through glycogenolysis. This mechanism of hormone action is known as the second messenger model. There is no need for the adrenaline to move through the cell surface membrane — it exerts its effect inside the cell using a transmembrane protein.

Another good explanation given at the start of the paragraph shows how hormones are relevant to both the theme and to earlier content.

Naming this enzyme shows very high-level knowledge.

Reference to liver cells reflects the title, which specifies the inclusion of different types of cell.

There is good emphasis here on the importance of membranes, and the student has avoided writing irrelevant content about the control of blood glucose levels.

The concluding paragraph draws together the topics discussed in the essay and emphasises why membranes are so important in different types of cell.

Membranes are important in cells and, although the general structure of the membrane is the same, different cells have different proteins on or within them so that they can carry out their function. Action potentials could not be created in neurones without ion channels and sodium-potassium pumps in their membranes, Pacinian corpuscles would not respond to mechanical stimuli without stretch-mediated Na^+ channels, and liver cells would not respond to adrenaline without specific receptors on their membrane surface. Within cells, proteins and other structures would be destroyed if enzymes, such as the hydrolytic ones found in lysosomes, could not be kept separate from the cytoplasm. Finally, mitochondria would not be able to produce the ATP essential for living processes without the proteins of the electron transfer chain located in their inner mitochondrial membranes.

This essay includes six detailed paragraphs (each of approximately 150 words), together with an introduction and conclusion, and is about 1300 words long. Although this seems a lot to write in 35 minutes, you should be able to write an essay of 1000–2000 words over about four or five pages within that time.

You should know

> Focus on the command word and make sure that your answer does what the question tells you to do.
> Use the correct scientific terminology throughout your answer.
> Remember that spelling, punctuation and grammar matter — proofread your work to check that it makes sense.
> Look carefully at the mark allocation and use this as a guide to how much you should write.
> Take the time to plan your answers to extended-response questions and essays rather than starting to write straight away.

4 Practical skills

Learning objectives

> To develop the thinking skills necessary to prepare for and understand practical activities

> To learn techniques that will help you work critically and purposefully during practical sessions

> To analyse, interpret and evaluate results after completing the practical work

> To develop the skills required to meet the standard for the practical endorsement

> To prepare for assessment of practical skills in written examinations

Every biology student will complete a minimum number of practicals so that they can use specified pieces of apparatus and demonstrate techniques. These practicals, together with the apparatus and techniques, are listed in the specification.

Practical skills are assessed in three ways during the A-level Biology course, depending on the exam board:

1 CPAC

2 Written papers

3 Practical exam

This chapter aims to support your practical activities over the 2-year A-level course by encouraging you to think more deeply and more critically about your practical work. There is guidance on each of the main parts of a practical: method, safety, results, conclusion and evaluation. The suggestions given will help you to meet the required standard for the practical endorsement and the practical exam. There are also worked examples of past paper questions that will prepare you for the assessment of practical skills in written examinations.

Practical assessment
1 Common practical assessment criteria (CPAC)

The majority of exam boards require students to demonstrate that they have met the five CPAC:

→ CPAC 1 — Follows written procedures

→ CPAC 2 — Applies investigative approaches and methods when using instruments and equipment

→ CPAC 3 — Safely uses a range of practical equipment and materials

→ CPAC 4 — Makes and records observations

→ CPAC 5 — Researches, references and reports

If you meet the criteria, and the majority of students are expected to meet them, then you will be awarded a 'Pass' for the practical endorsement on your A-level certificate. Many universities will ask for a 'Pass' from students as one of the entry requirements, even if you are not applying for a science course.

The assessment and monitoring of the CPAC will vary slightly between exam boards, and between schools or colleges. Some students may work in pre-printed practical booklets, whereas others may record their practical work in a hard-backed notebook. This chapter will refer to a **lab book** as the place in which you record your practical results. The skills and the standard required are the same for all students, despite any differences in their assessment or monitoring.

2 Written papers

Many of the written papers will have questions assessing the practical skills. These practical skills exam questions will include questions on both the theory and application of practical skills, and will account for *at least 15%* of the overall mark (as specified by Ofqual). Some exam boards have all of the practical skills questions in a written practical paper, whereas other exam boards assess the practical skills in every exam paper.

Skills assessed in the written papers are as follows:

→ Independent thinking — solving problems and applying your scientific knowledge.

→ Use and application of scientific methods and practices — identifying variables, presenting data, evaluating methods and results, drawing conclusions.

→ Numeracy and the application of mathematical concepts in a practical context — plotting and interpreting graphs, processing and analysing data, considering quality of data (margins of error, accuracy and precision).

→ Instruments and equipment — knowing how to use, and understanding, a wide range of apparatus and techniques.

 Exam tip

You must revise the practical content as part of your exam preparation. Make sure that you know the format of the exam papers so that you know *when* to revise the practical work.

3 Practical exam

CCEA and WJEC include a practical exam.

For CCEA, teachers assess student performance in a range of practical tasks. The skills assessed include:

→ carrying out the methodology specific to the practical task

→ collecting data while managing any associated tasks

→ recording data in tables and/or graphs

→ drawing evidence-based conclusions

→ demonstrating other skills appropriate to individual practical tasks, for example the completion of block diagrams

Students complete at least seven practicals at AS and at least five at A2. Each practical is assessed on a 3-point scale, with a zero score awarded to students who fail to attempt the task, and a score of 3 to 'candidates who show **competence**, **independence** and **skill** in fully completing and fully recording the practical task'.

For AS Unit 3, 21 marks out of a total of 71 are from this teacher assessment (a maximum of 3 marks for each of the seven practicals), and 15 out of 75 marks are available from teacher assessment in A2 Unit 3.

For WJEC, A2 Unit 5 is the practical examination, which has two tasks: an experimental task (20 marks) and a 60-minute practical analysis task (30 marks).

The experimental task will consist of one 2-hour session. You will be provided with a set of apparatus and an examination paper containing a method to follow — for example, investigating the effect of temperature on the rate of carbon dioxide production in yeast. Your teacher will directly assess some of the practical skills — for example, is the apparatus set-up correct? You will then record your results in a suitable table, plot a graph of your results and answer questions about the technique. These questions could be about sources of inaccuracy, or modifications to the procedure.

Study skills

Method — following written procedures

Preparation before the practical

Read the method before starting a practical activity. This seems like an obvious thing to state, but many students launch into the practical without any real idea of what they are doing and why. Ideally, read through the method before you arrive in the laboratory so that you can maximise your laboratory time.

Imagine yourself carrying out the practical as you read the method. You may find it useful to draw diagrams or write additional notes to support your understanding of the procedure. For example, if you are setting up several test tubes with different contents then you could sketch the test tubes and annotate them to show the contents.

Plan the most efficient way of organising your equipment and completing the practical.

Think carefully about *why* you are carrying out each step rather than following procedures unquestioningly.

Revise or research the background information — for example, how a colorimeter works, or why qualitative reagents change colour.

During the practical

Following written instructions is evidence for CPAC 1, essential during a practical exam, and an important step towards independence in the lab. Do not just copy what other students are doing. If you are using an unfamiliar technique, you should have the written procedure in front of you during the practical so that you can follow it methodically and avoid missing steps.

Your teacher will assess you during the practical sessions to make sure that you are meeting the required standard. This may be through observation or by asking you questions about the techniques.

If you have any waiting time during a practical — for example, while tubes of solution reach the temperature of the water bath — make good use of it:

→ clear away any equipment that you have finished with
→ keep your work area tidy and organised
→ lay out equipment in the order in which you will use it
→ set up the next set of tubes if you are doing repeats

By the end of the A-level course, you should be able to work confidently and demonstrate **fluency** with familiar apparatus and techniques.

Worked example 4.1

Some exam questions assess your familiarisation with the practical techniques listed in the specification. You will not be expected to recall an entire practical procedure, but in the exam you may be given information about a practical activity and then asked what the next steps would be. These questions often include a list of steps that a student has already followed, or a list of equipment that you need to use, and then ask you to describe the method.

A student investigated the pigments present in a leaf. The student:
→ **crushed a leaf with solvent to extract the pigments**
→ **drew a pencil line 1 cm from the bottom of a strip of filter paper**
→ **used a capillary tube to add a spot of pigment extract to the pencil line**
→ **stood the strip of filter paper in a boiling tube containing solvent**
→ **removed the filter paper from the boiling tube before the solvent reached the top**

Describe how the student could continue the experiment to find the Rf values of a leaf pigment. (4)

This is a straightforward 'describe' question, but it is worth quite a lot of marks. You are given a lot of information and guidance in the question, so read it carefully before you start writing. The main thing to remember is that your answer must be structured logically, with a clear sequence of steps.

The student should draw a pencil line to mark the solvent front ✓, then use a ruler to measure the distance between the origin (the first pencil line) and the solvent front ✓. The student should then find the distance between the origin and the centre of the pigment spot ✓ and divide this distance by the distance moved by the solvent ✓.

The student has written a clear sequence of steps that would allow calculation of the Rf value.

Method — applying investigative approaches

Preparation before the practical

Your teacher may ask you to plan a full investigation, with the following:

→ Hypothesis — a testable statement, for example, 'the wavelength of light will affect the volume of gas produced by pondweed'.

→ Equipment list — all of the equipment required for the practical specified, with justification of choices where appropriate.

→ Variables — independent, dependent and control variables all identified. Explain *how* variables will be controlled and *why* they need to be controlled.

→ Method — a clear set of instructions that could be followed by someone else.

During the practical

You should be able to select appropriate pieces of equipment and to justify your choice. For example, if you needed to measure $1\,cm^3$ of solution, choosing a $1\,cm^3$ syringe would have a smaller uncertainty than a $5\,cm^3$ syringe, or a $10\,cm^3$ measuring cylinder. (The uncertainty of a measurement is plus or minus half the smallest scale division.)

A student demonstrating a high level of practical competency checks the steps and makes adjustments if necessary. For example, if you are told to leave test tubes of solution in a water bath for 5 minutes to equilibrate then you may question whether this is the optimum length of time. By using a thermometer to check the temperature of the solutions inside the test tubes, you may determine that 5 minutes is insufficient time and then modify the procedure.

If you are not getting the results that you expect, then troubleshoot the problem yourself rather than rely on the teacher. Double check the procedure to make sure that you have not omitted a step; repeat the procedure using a different piece of equipment or different source of reagent. You might double check the temperature of the water bath and find that it is actually 42°C rather than 40°C, and then change this in your results table. This will show that you fully understand the activity that you are carrying out and are able to link your actions to the expected outcome of the investigation. If you make any modifications, or spot any potential issues during the practical, make a note of these in your lab book, together with an explanation of why you made the modification.

Worked example 4.2

Exam questions assessing your understanding of the investigative process may require you to apply both your practical skills and understanding of the specification to an unfamiliar example. These questions often include reference to control variables, control experiments and the collection of valid data.

A student investigated the light-dependent reactions of photosynthesis using a suspension of isolated chloroplasts and a blue dye called DCPIP. DCPIP loses its blue colour when it is reduced.

The student set up three tubes:

Tube 1 — chloroplast suspension + DCPIP

Tube 2 — chloroplast suspension + DCPIP

Tube 3 — solution without chloroplasts + DCPIP

The student put Tube 1 in the dark for 30 minutes, and Tubes 2 and 3 in the light for 30 minutes.

After 30 minutes, the student recorded the colour of each tube.

(a) The student made the chloroplast suspension using a solution with the same water potential as the chloroplasts. Explain why. (2)

(b) Explain why the student set up tube 3. (2)

(c) Describe and explain the appearance of the three tubes after 30 minutes. (3)

You should have the background knowledge about the light-dependent reactions and then make the link that the DCPIP would be reduced if it gained electrons released from chlorophyll during photoionisation. You also need the practical knowledge of the purpose of a control experiment — tube 3 has been set up as a control. Finally, you need to apply your knowledge of water potential and its effects on cells to the effect it would have on isolated chloroplasts.

(a) You need to apply your knowledge of the movement of water into and out of cells to the effect water potential would have on chloroplasts.

Step 1: Show your understanding of the term 'water potential' and correctly use the term 'osmosis'.

Step 2: State what would happen if the water potential was not the same as that of the chloroplasts.

The solution has the same water potential as the chloroplasts, so that water does not enter or leave the chloroplasts by osmosis. ✓ *If water entered the chloroplasts then they might burst/if water left the chloroplasts then they might shrivel.* ✓

Note here that the student has referred to water entering or leaving *the chloroplasts* and them bursting/shrinking. A common error would be for students to forget that the chloroplasts have been isolated and to refer to movement of water into/out of *cells*.

(b) This question is assessing your understanding of control experiments. Note that there are 2 marks available for this question, so you would have to give *two* clear reasons why the control tube was set up.

Tube 3 has been set up as a control to show that chloroplasts are needed for decolourisation of the DCPIP ✓ *and to show that the DCPIP does not decolourise due to the presence of light* ✓.

(c) This question has two command words: 'describe' and 'explain'. For each of the three tubes, you need to make sure that you *describe* the appearance (blue or colourless) *and explain* why this would happen. Your explanation needs to demonstrate your understanding of both the reduction of DCPIP and the release of electrons during the light-dependent reactions. Present your answer clearly and logically, with reference to each tube in turn.

Tube 1 will stay blue because no light is present to excite the electrons from chlorophyll, so the DCPIP is not reduced and retains its blue colour. ✓

Tube 2 will change colour from blue to colourless because the light will excite electrons from the chlorophyll in the chloroplasts and these electrons will be accepted by DCPIP and change it to the colourless reduced form. ✓

Tube 3 will remain blue because it contains no chloroplasts (and no chlorophyll) to provide electrons, so the DCPIP is not reduced. ✓

In this answer, the student has made it clear which tube they are referring to, and for each tube has specified the colour and given the reason for the colour.

Take it further

The Ten Most Beautiful Experiments, by George Johnson, describes the work of some of the greatest scientists and the curiosity that led them to incredible discoveries. It includes Harvey's research into the circulation of blood, Galvani's discovery that there was a link between electricity and muscle contraction and Pavlov's famous research using dogs that would salivate in response to sound, and provides a fascinating insight into the scientific process.

Safety

Preparation before the practical

You may be asked to complete a risk assessment before taking part in the practical. A risk assessment includes the identification of possible hazards, the risks associated with these hazards, and the control measures that you will take to minimise the risk of injury. Having read through the method you may choose to research the risks associated with specific chemicals or techniques. For example, if you were completing a dissection, then your risk assessment could include the information shown in Table 4.1.

Table 4.1 Part of a risk assessment for a dissection

Hazard	Risk	Control measures
Scalpel	Cutting yourself	Cut away from yourself
		Use a sharp blade that has no rust
		Cut onto a suitable surface

Activity

A student is about to complete a practical to show that sucrose is a non-reducing sugar. The student:

- adds $2\,cm^3$ of sucrose solution and $1\,cm^3$ of hydrochloric acid to a test tube
- boils the sucrose solution and hydrochloric acid for 1 minute
- neutralises with sodium hydrogen carbonate
- adds $3\,cm^3$ of Benedict's solution to the test tube
- stands the test tube in a beaker of boiling water for 5 minutes

Write a full risk assessment for this practical activity.

During the practical

Your teacher may assess your ability to work safely without prompting. You should be aware of hazards and risks, and work in a confident and sensible manner. The expected behaviours would include:

→ glassware positioned away from the edge of the bench
→ spills wiped up promptly with a paper towel
→ stoppers replaced on bottles of reagent
→ accidents and breakages reported to the teacher

A student demonstrating a high level of competency in the laboratory would be aware of all possible risks and would take appropriate measures to control them. They would also take responsibility for the safety and wellbeing of other students, dealing with incidents calmly and appropriately.

Results

Recording observations

When preparing for a practical activity, you need to consider how you will record your results. In biology this often takes the form of a biological drawing or a results table. Your results must be recorded neatly and methodically in your lab book. You must also make sure that your observations are *accurate*.

Biological drawings

For biological drawings, you must ensure that you:

→ draw a large enough diagram to show detail

→ use a sharp pencil

→ draw single, clear lines, and do not sketch

→ do not colour or shade

→ draw to scale (with the scale given if appropriate)

The difference between...

Figure 4.1 shows a photomicrograph of root-tip cells undergoing mitosis.

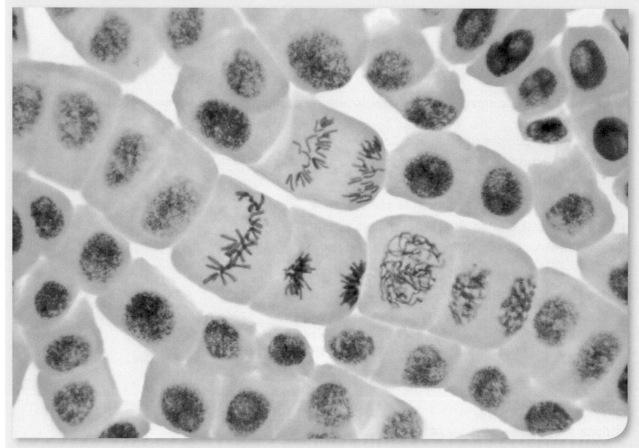

Figure 4.1 Cells from the tip of a plant root

This has been drawn by two students (Figures 4.2 and 4.3):

The difference between…

A-grade response	C-grade response
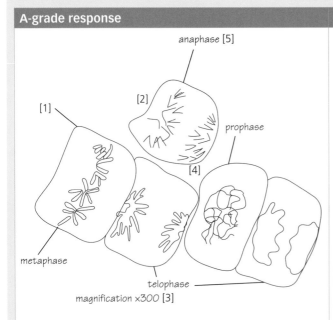	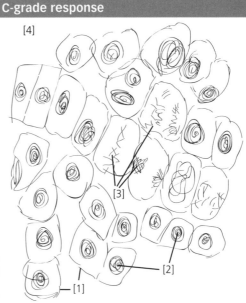

Figure 4.2

Figure 4.3

[1] *Single, clear lines have been drawn using a sharp pencil.*

[2] *There is no shading or colour.*

[3] *The magnification has been included.*

[4] *The student has only drawn five cells, but has drawn them accurately.*

[5] *The stages of mitosis have been identified.*

[1] *The student has used broken, sketched lines.*

[2] *Biological drawings should not have any shading.*

[3] *There is a lack of detail, with chromosomes roughly sketched.*

[4] *The student has tried to draw all of the cells rather than drawing a few cells accurately.*

Tables

Draw a **results table** straight into your lab book or wherever you are recording your data. Do not scribble results onto scraps of paper that you might lose. Even if your data are raw data that you then want to record neatly, you should still have a record of these raw data in your lab book.

Decide how many repeats you will take so that you can include an appropriate number of columns. Plan ahead to how you might process your results — is a further column required for rate or percentage change?

As you collect your results, record them straight into your results table. For example, if you are using a balance to record the mass, take your lab book over to the balance with you and record the mass directly into your lab book.

Process your results as you collect them and sketch a graph if appropriate. This will help you to identify anomalies and deal with them straight away. Do not just ignore anomalous results. If any of the results appear anomalous, then you should repeat the reading. If the repeat is still anomalous, try to consider *why* it might be anomalous. For example, if you are using a thermostatically controlled water bath, double check the temperature using a thermometer.

Annotated example

Results tables should have the following:

The independent variable in the first column

Full column headings rather than just, for example, 'Time'

No mixed units, for example, convert into seconds rather than both minutes *and* seconds

Units in the column heading only and not in the body of the table

Distance between lamp and chloroplast extract/cm	Time taken for the DCPIP to decolourise/s				Rate of decolourisation/s⁻¹
	Reading 1	Reading 2	Reading 3	Mean	
5					
10					
15					
20					
25					

Activity

A student investigated the effect of temperature on the rate of a reaction catalysed by the enzyme trypsin. The student used a dilute solution of milk powder containing the protein casein. When trypsin hydrolyses casein, the milk powder solution becomes clear.

For this investigation:

(a) Identify the variables that would need to be controlled, then explain how and why they should be controlled.

(b) Decide on a suitable range for the independent variable and explain the reasons for your choice of range.

(c) Draw a results table for the investigation, remembering to include columns for repeats and processed data.

The practical exams for CCEA and WJEC award up to 5 marks for the following:

1 Setting up your apparatus (without help from the teacher).

2 Selecting an appropriate range for the independent variable, or selecting an appropriate value for a control variable — for example, the temperature for an enzyme-controlled reaction.

3 Results table drawn, with correct column headings.

4 Suitable units in column headings only.

5 Data processed correctly, for example, means and/or rates calculated.

These are straightforward marks to obtain, so developing good habits throughout the course will ensure that you always produce a suitable results table.

 Exam tip

When completing a table to an 'appropriate number' of decimal places, look at the existing values in the table and give your answer to the same number of decimal places.

Quality of data

You are expected to obtain **accurate**, **precise** and **sufficient** data during a practical. You may also be asked to assess the quality of data in the written exams.

Accurate data

Accurate means that the measurement is close to the true value. In practice, the results that you collect should be close to those obtained by your teacher or technician, or close to the known value. For example, if you are investigating the effect of pH on a specific enzyme, the teacher or technician will have information about the expected optimum pH of the enzyme and your results will be judged accurate if they are close to the expected value.

Precise data

Precise means that there is little spread around the mean value, or a smaller uncertainty. For example, if human body temperature is accepted as 37.0°C, then an experiment that gives a body temperature of 35.5 ± 0.1°C is *more precise* than an experiment that give 37.0 ± 1.0°C, but the latter is *more accurate*.

The term precision is also used when referring to a measurement scale — for example, a thermometer might measure to a precision of one degree (± 1.0°C).

Sufficient data

Decide for yourself when sufficient data have been collected. This is not always three sets of repeats. If you repeat the procedure and obtain a second set of concordant results, then there is no need to collect a third set. If you obtain three sets of data showing variability, then you may need to repeat a fourth or even a fifth time.

Make sure that you know how many repeats you need to do for statistical analysis because there may be a minimum number of results needed for the statistical test. For example, if you were using Spearman's rank as a test of correlation, you would need to collect between 10 and 30 paired observations. It may not be possible to go back and collect more results when you realise that you do not have enough data for a statistical test.

> ### ✓ Exam tip
>
> Compound units are made up of two or more units and should be expressed using negative indices. For example, cm^3 per second should be written in the form $cm^3 s^{-1}$ rather than cm^3/s.

Worked example 4.3

At the planning stage of an investigation, you need to decide on the quantity of data to collect. Exam questions may require you to demonstrate your understanding of data collection and processing, particularly with regard to the use of statistical tests.

A student investigated the effect of air pollution on seed size in one species of tree. The student:
- → **collected seeds from trees in a city**
- → **collected seeds from the same species of tree in the countryside**
- → **processed the data to determine if the seed size was different**

Describe how the student could collect the seeds and process the data to determine if there was a difference in seed size. **(5)**

➡

Step 1: Describe how the student will collect valid data.

Although you are not expected to have completed this investigation, you should be able to apply your knowledge of sampling techniques to collect sufficient and valid results.

The sampling must be *random* to avoid sampling bias, so the technique you describe should explain how this is achieved.

The sample size must be as *large* as practically possible to minimise the effect of anomalies when calculating the mean. A larger sample size will be more representative of the population as a whole.

You need to specify what you will actually measure to determine 'seed size' — for example, you could measure mass, length or circumference.

> The student needs a large number of seeds from each population of trees and should collect 100 seeds from trees in the city centre and 100 seeds from trees in the countryside. ✓ A random sampling technique should be used when collecting the seeds to reduce sampling bias ✓, and the student should collect the seeds from the same number of trees in each location. Once the seeds have been collected, they should be stored in airtight containers to minimise water loss, and the mass of each seed should be recorded. ✓

This answer makes clear reference to the sample size and importance of random sampling, but also realises that there could be variation between different trees within each population. By collecting the seeds from different trees within each population, natural variation is taken into account.

Step 2: Describe how the student will process the results to see if the difference is significant.

The question asks you to *process* the data and describe how the student would find out if there were any *difference* between the seed sizes. This suggests that statistical analysis is required. Your answer needs to be specific and should explain exactly *how* you would process the data and which statistical test you would use.

> After finding the mass of seeds at the two different locations, the student should calculate the mean and the standard deviation ✓ for each tree population. A statistical analysis should be carried out to see if there is a significant difference ✓ between the mean seed masses from the two different locations. The Student t-test is used to judge the significance of any difference between the means of two sets of data. ✓

Processing data

After the practical session, the data need to be processed and analysed and then any patterns or trends can be identified. The reasons for the trends should then be explained using your scientific knowledge when you write your conclusion.

You should be able to process your data in different ways, including the use of software and other tools. Processing your results may involve:

→ calculations, for example, mean, standard deviation, rate of reaction
→ statistical analysis
→ plotting graphs

Some guidance on calculations and statistical analysis is given in chapter 1 (Quantitative skills), but you also need to refer to the specification for mathematical skills when processing data.

 **Common pitfall**

Remember that you should use all of the data given when calculating the mean unless you are specifically told to disregard a number. Do not assume that results should be omitted, even if they seem to be anomalous.

Plotting graphs

Plotting a graph of your results allows patterns to be seen clearly and is also an important skill that will be assessed in the written examinations. It should be a straightforward task, but students often lose marks for making careless mistakes. Some exam boards award up to 9 marks for plotting a graph, so it is worth taking the time to practise this skill.

Refer to the following checklist to ensure that you get full marks:

➜ Size — use at least half of the available space.

➜ Axes — must be the correct way round, with the independent variable on the *x*-axis.

➜ Scale — must be suitable and allow for accurate plotting (equal and sensibly chosen increments).

➜ Labels — axes must be fully labelled with a description of the variable *and* the units.

➜ Plotting — points should be plotted accurately with a dot or a cross.

➜ Curve — points should either be joined with straight lines or a smooth curve of best fit drawn as appropriate.

➜ Error bars (or range bars) — should be included if necessary.

 Exam tip

When plotting a graph, choose a scale that is easy to work with. Increasing in increments of 1, 2, 5, 10 or 100 will give a much easier scale to use than increments of, say, 3, 7 or 9.

Activity

Use the data in Table 4.2 to plot a graph. Make sure that you have followed all of the guidance given.

Table 4.2 Sample data for plotting

pH	Rate of reaction/$g\,cm^{-3}\,min^{-1}$
4	0
5	1
6	1
7	3
8	6
9	5
10	2
11	0

The difference between...

Figures 4.4 and 4.5 show graphs of the data in Table 4.2 drawn by two students:

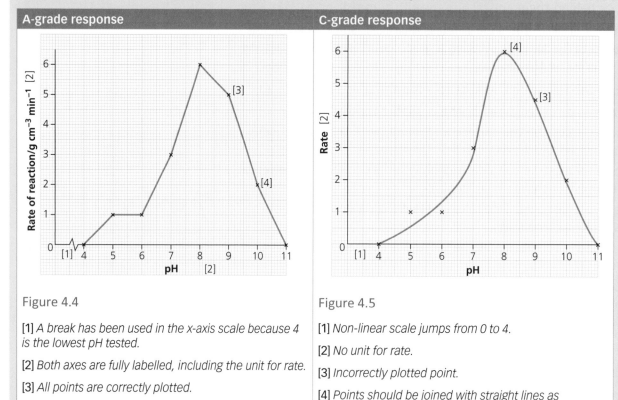

A-grade response	C-grade response

Figure 4.4

Figure 4.5

[1] *A break has been used in the x-axis scale because 4 is the lowest pH tested.*

[2] *Both axes are fully labelled, including the unit for rate.*

[3] *All points are correctly plotted.*

[4] *Points are joined with straight, ruled lines.*

[1] *Non-linear scale jumps from 0 to 4.*

[2] *No unit for rate.*

[3] *Incorrectly plotted point.*

[4] *Points should be joined with straight lines as maximum rate could be at a pH not measured, e.g. 8.5.*

Conclusion

Reporting the findings of your investigation requires you to analyse your results and then draw appropriate conclusions. When explaining the results of the investigation, you should include a high level of detail. Use **research** to support your findings, and remember to **reference** all sources.

✓ **Exam tip**

Correlation between two variables does not necessarily mean that there is a causal link. Further investigation may be required to show that the two variables are linked.

Reporting

After you have plotted a graph, look at the overall shape of the graph and describe the trend it shows. The trend should refer to both variables and the pattern shown by the graph — for example, 'as the light intensity increased, the volume of gas collected from the plant increased'. Your trend should also refer to the data — for example, 'the rate increased rapidly between 0 and 3 minutes, then levelled off at a rate of 17 cm^3 s^{-1}'.

Once you have identified the trends, you need to use your scientific knowledge to explain them. Explaining your results gives you an excellent opportunity to revise the theory that you have covered in class and make links between your theory and practical work.

Research and referencing

Research requires you to read relevant sources and then use the information to complete your written report *in your own words*.

Before completing an investigation, you could research two or three different methods, evaluate them, and then decide which one to use. After an investigation, you could read about and research the science behind the practical and use this information to give a full analysis of your results.

Remember that all sources must be fully referenced, with the date of access for websites as well as the full URL. You may also be asked to use a specific referencing system, such as the Harvard referencing system.

Worked example 4.4

In the exam you may be given information about a practical technique and asked how you could *modify* or *adapt* a procedure to investigate a different factor. Some exam questions may take this further and ask how you can *use your results* or *draw a conclusion*.

Your answer to this type of question needs to be structured logically and must include everything that you are asked to do. For example, if you are asked to describe a method and explain how you would use your results, these are two separate things that must be specifically addressed.

Figure 4.6 shows a potometer.

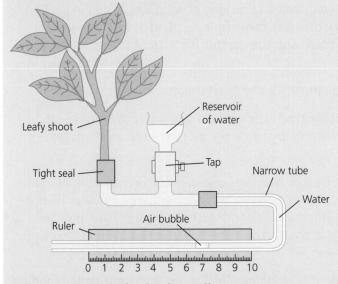

Figure 4.6 A typical school or college potometer

A student uses a potometer to investigate the rate of water uptake by a leafy shoot.

The student:
- **cuts the leafy shoot and places it in the potometer under water**
- **rubs petroleum jelly around the seal to prevent air from entering and water from leaving**
- **waits 5 minutes for the leafy shoot to adjust**
- **records the position of the air bubble and starts a stop watch**
- **records the position of the air bubble after 10 minutes**

Describe how this method could be modified to identify the leaf surface with the greatest number of stomata, and explain how the results of the investigation could be used to reach a conclusion.

(5)

Step 1: Describe the method.

The method does not need to be long and detailed. The total number of marks available for the question is 5, so you should not need to write more than three sentences for the method.

If you are unfamiliar with the use of a potometer, look carefully at the diagram and read the method the student used to set up the equipment. The use of petroleum jelly is a clue — if it prevents water from leaving the equipment, it will also prevent water from leaving the leafy shoot via the stomata.

Do not repeat information given in the stem of the question. For example, no marks are available for stating that the leafy shoot will be cut under water. Just describe how you would *modify* the method.

> Cover the upper surface of the leaves with petroleum jelly and record the distance moved by the air bubble in 10 minutes ✓, then repeat using a different leafy shoot with the lower surface of the leaves covered with petroleum jelly ✓. The leafy shoots should be from the same type of plant, the surface area of the leaves needs to be the same, and any environmental conditions that might affect water loss from the leaves should be kept constant, for example, air movement and temperature. ✓

This is a clearly written answer, which makes very good reference to control variables and will ensure the collection of valid results.

Step 2: Use the results to reach a conclusion.

When explaining how you would use the results, you need a clear idea of what would you expect to happen and why. The key idea is that the majority of water taken up by a plant is lost through the stomata during transpiration. The more stomata there are, the greater the rate of transpiration and therefore the higher the rate of water uptake.

Include comparative terms in your answer, such as higher/lower or less/more.

Remember that the potometer makes the assumption that water uptake and water loss are equal. This is not the case because some of the water taken up by the leafy shoot is used as a reactant in photosynthesis, or for other cellular processes.

> Compare the rate of water uptake in the leafy shoot with the upper leaf surface covered with petroleum jelly, and the shoot with the lower leaf surface covered. Assuming that water uptake and water loss are equal, the shoot that has the highest rate of water uptake is the one with the highest rate of water loss. ✓ As this water loss takes place through the stomata, the highest rate of water loss is through the surface with the greatest number of stomata. ✓

This answer is very well structured, with good use of terminology, and shows excellent understanding of the procedure. The answer is well balanced with a clear description of the procedure *and* an explanation of how to use the results to reach a conclusion.

Evaluation

The two main features of an A-grade student's post-practical analysis are the level of detail and the level of criticism. You need to critically evaluate the procedure you used, the quality of your data and the validity of the conclusions you have reached.

Use the correct scientific terminology when you are evaluating your investigation rather than using terms such as accurate, precise and reliable interchangeably.

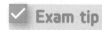 **Exam tip**

Accurate refers to measurements that are close to the true value. Measurements are **precise** if there is very little spread about the mean. It is possible for measurements to be precise, but not accurate.

Developing these evaluative skills is important for two reasons. Firstly, it will make you a better researcher who is able to work independently and creatively. Secondly, and this is of more immediate benefit to you, these skills are assessed in the written examinations.

Table 4.3 gives examples of types of question that assess your ability to evaluate investigations.

Table 4.3 Practical evaluation questions

Example of question	Explanation
'Suggest a **limitation** of the procedure...'	Look at the procedure critically. Is there anything that the student/scientist has done or omitted to do that will affect the validity of the results — for example, has a method for random sampling been used to avoid sampling bias?
'Discuss the **correlation** between...'	Look for a relationship between the two variables. Is there positive/negative/ no correlation? If there is a correlation, then write a statement linking the two variables together — for example, 'as the concentration of fertiliser increases, the mass of the seedlings increases'. Suggest reasons why there may or may not be a correlation.
'How could the **accuracy** of the data be improved?'	Look for ways in which the results could be closer to the true value. This could involve the use of a different piece of equipment, or an increase in the sample size to make the mean value more representative of the whole population.
'Discuss the **validity** of the conclusions...'	Look for ways in which the procedure and results do or do not support the conclusions. For example, a procedure that uses a large sample size or a random sampling technique to reduce sampling bias would produce valid data from which valid conclusions can be drawn. A procedure that relies on subjective judgements or human interpretations of the measurements could produce questionable data and doubts about the validity of the conclusions drawn from the data.

Worked example 4.5

Questions assessing higher-order thinking skills may require you to evaluate procedures, results and conclusions. Some questions will outline a procedure and assess your ability to identify the strengths and/or weaknesses of the investigation.

Students investigated the effect of the Bt toxin on insect pests that feed on maize plants.

The students:

1 divided a field into three plots
2 planted hundreds of maize seedlings in each plot:
    → Plot 1 — untreated seedlings
    → Plot 2 — seedlings genetically modified to produce Bt toxin that kills insect pests
    → Plot 3 — seedlings sprayed with Bt toxin every week
3 counted the number of seedlings that had died each week. Each student counted a different plot.

The students' teacher was concerned about the investigation and suggested that:
→ **the method may not be valid**
→ **not all variables have been controlled**
→ **the results may not be accurate**

Explain why the teacher's concerns are justified and suggest improvements to the investigation.   (6)

You are told that the teacher has concerns and you have to identify the causes of these concerns. The use of the term 'justified' means that the teacher was right to be concerned, so your answer needs to find the flaws with the investigation.

This question requires you to respond to a lot of information, but gives you very specific guidance on what to include. A question like this may be marked using level-of-response marking, in which you have to include:
→ a critical evaluation of the procedure
→ a clear and logically structured answer
→ a well-developed line of reasoning
→ relevant, substantiated information

Step 1: Explain why the method may not be valid *and* suggest improvements.

To collect valid data, your measurements must only be affected by a single, independent variable. Control variables must be kept constant and there should be no bias.

The question wants you to focus on potential flaws in the investigation, but your suggestions need to be specific rather than vague references such as 'it is not a fair test'. The question does not give the mark allocation for each part of the answer, but try to include at least two reasons why the method may not be valid.

The students counted the number of seedlings that had died, but they may have died for another reason. The students could look for evidence that seedling death was caused by insect pests — for example, are there holes in the leaves? ✓

The method does not state the initial number of seedlings, so recording the number of seedlings that have died in each plot does not allow the plots to be compared. The students should count the initial number of seedlings in each plot and record the deaths as a percentage. ✓

Step 2: State any variables that have not been controlled *and* suggest improvements.

There is overlap between this step and step 1, because in order to collect valid results, you need to control variables that could affect the dependent variable. The stem of the question refers to the plural 'variables', so try to think of at least two suggestions.

There may be genetic variation between the different seedlings that could make them more susceptible to the insect pests. The students could use cloned plants in each plot. ✓

The students did not control the number of seedlings in each plot. Seedlings that are planted more closely together may allow pests to spread more easily between plants. The students should plant the same number of seedlings the same distance apart in each plot. ✓

Step 3: Explain why the results may not be accurate *and* suggest improvements.

Remember that accuracy refers to how close a result is to the true value. You should be considering any issues with the method that may cause the students to miscount the true number of dead seedlings.

The method does not state the size of the plot, but as it is a whole field divided into three plots with hundreds of seedlings, this suggests that each plot has a large area. As there are 'hundreds of seedlings' and the students are walking around the edge of the plot to count them, it would be difficult to see seedlings in the centre of the plot. An improvement to the method needs to allow a more methodical approach to planting the seedlings and counting the dead ones.

The students may not see dead seedlings in the centre of the plot. The seedlings could be planted in rows so that the students can walk up and down each row and see every seedling clearly. ✓

Different students counting each plot may interpret what is meant by a 'dead seedling' differently. Each student could count all three plots and then a mean could be calculated. ✓

Application to the exam

This section includes exam questions assessing some of the skills discussed in this chapter. The questions both refer to the respirometer, but require you to demonstrate different skills.

After each question, there is guidance on how to approach the question and how to structure your response. For each question:

→ read the guidance carefully

→ answer the question under timed conditions

→ self-assess your answer using the mark scheme

→ compare your answer with the sample answer

→ rewrite your answer, if necessary, to produce a model answer

Exam question 1

Figure 4.7 shows a respirometer.

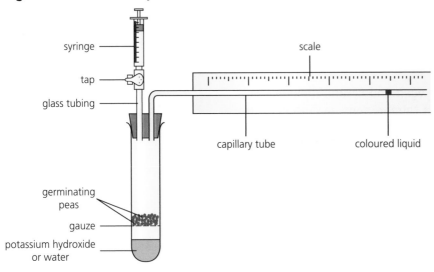

Figure 4.7 Respirometer

A student used the respirometer shown in Figure 4.7 to investigate anaerobic respiration in germinating peas. Anaerobic respiration in plants produces carbon dioxide. The student covered the tube with foil to prevent photosynthesis.

Devise a plan to determine if the germinating peas are respiring anaerobically. **(6)**

Your plan should include:

→ an outline of the experimental set-up (you do not need to give a detailed procedure)

→ variables to control

→ data to collect

How to approach the question

This question requires your understanding of respiration and of the general principles of the respirometer. Even if you are unfamiliar with the equipment, there is enough information given in the stem of the question and the diagram for you to apply your knowledge. The clue in the diagram is the label 'potassium hydroxide or water'.

If the respirometer was set up *with* potassium hydroxide, any carbon dioxide produced by a respiring organism would be absorbed. *Aerobic respiration* would remove oxygen from the air in the respirometer, but this is not replaced by carbon dioxide as the carbon dioxide is absorbed. This causes a decrease in pressure inside the respirometer. As the pressure inside is now lower than atmospheric pressure, the drop of coloured liquid in the capillary tube *moves towards* the respiring organisms. *Anaerobic respiration* would also produce carbon dioxide, but oxygen would not be removed from the air in the respirometer. The carbon dioxide would be absorbed by the potassium hydroxide, so the pressure inside the respirometer would not change and the drop of coloured liquid would *not move*.

If the respirometer was set up *without* potassium hydroxide and had water instead, the carbon dioxide would not be absorbed. *Aerobic respiration* would remove oxygen from the air inside the respirometer and replace it with carbon dioxide. The pressure inside the respirometer would remain constant and the drop of coloured liquid would *not move*. *Anaerobic respiration* would produce carbon dioxide and the pressure inside the respirometer would increase. The drop of coloured liquid would *move away from* the respiring organisms.

The effects of respiration on the oxygen and carbon dioxide levels inside the respirometer described in the previous paragraphs will change if photosynthesis is also occurring. Photosynthesis must be prevented by covering the tube when investigating respiration.

Your plan needs to include a respirometer set up both with potassium hydroxide and without potassium hydroxide. The drop of coloured liquid not moving with potassium hydroxide and moving away from the organisms with water will demonstrate that anaerobic respiration is taking place.

This exam question gives you a lot of guidance on how to structure your answer. Plan your answer using the points given and make it explicit that you have addressed each point. For example, you could start your sentences with 'The variables to control are...' and 'If the germinating peas are respiring anaerobically...'.

Give yourself 10 minutes to answer this exam question.

Mark scheme

1 respirometer used with potassium hydroxide and then with water;
2 temperature controlled using water bath;
3 same peas/same number of peas/same mass of peas;
4 distance moved by coloured drop measured in set time;
5 (with potassium hydroxide) coloured drop will move towards peas if aerobic/if oxygen uptake is greater than carbon dioxide production;
6 (with water) coloured drop will not move if aerobic/if oxygen uptake equals carbon dioxide production, but will move away from peas if anaerobic/if carbon dioxide production is greater than oxygen uptake;

Marking point 4 is awarded here. The question asks you to specify the data that will be collected, and this well-written answer would allow the calculation of rate. The student has also referred to the direction of movement, which is an important point to note.

There is now enough to award marking point 6 because the student has explained what would happen with both aerobic and anaerobic respiration if water was used in the respirometer. This final paragraph links clearly to the question, with an explanation of how the respirometer can be used to show that the peas are respiring anaerobically.

Sample answer

The peas are put into the tube with potassium hydroxide and the tube is sealed. The start position of the coloured liquid is noted and a stop watch is started. The direction of movement and the distance moved by the coloured liquid in 10 minutes is recorded. ✓ The potassium hydroxide is removed and replaced with water, then the movement of the coloured liquid is measured again for 10 minutes. ✓

Variables to control include temperature, as this affects the rate of respiration. Temperature can be controlled using a water bath. ✓ When doing repeats, the same peas should be used each time because peas at different stages of germination could have different rates of respiration. ✓

If the peas are respiring aerobically the drop of liquid will move to the left (towards the peas) with potassium hydroxide and will not move with water. ✓ If the peas are respiring anaerobically, the drop will move away from the peas when water is used ✓ because the production of carbon dioxide will increase the pressure in the respirometer.

Marking point 1 is awarded here. The stem of the question asks for 'an outline of the experimental set-up' and this brief but clear description includes sufficient detail.

The student has stated which variables need to be controlled, and suggested how and why they should be controlled. Marking points 2 and 3 are awarded for this answer.

Marking point 5 is awarded here and this sentence also includes the first part of marking point 6.

Exam question 2

Describe a procedure using a respirometer to compare the rate of respiration of a mouse with the rate of respiration of a woodlouse.

Your answer should include:
→ **explanations of modifications to the procedure**
→ **ethical treatment of the animals**
→ **collection of valid results**

Comment on the results you would expect and the conclusions you would make. **(6)**

How to approach the question

This question has lots of different sections to it and requires careful planning to meet all of the marking points.

You need to do the following:

1 Describe a procedure that will allow you to compare the rates of respiration in the different animals.

2 Explain any *modifications* — the command word here is 'explain', so as well as stating what the modifications are you need to say *why* they are necessary.

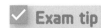
✓ Exam tip

When an exam question has a list of things that you need to include in your answer, tick off each one as you complete it so that you do not miss anything out.

3 Describe the *ethical treatment* of animals — this refers to the precautions you would take to minimise harm or distress to the animals. Potassium hydroxide is an irritant at concentrations of 0.5–2.0% and is corrosive at concentrations greater than 2.0%.

4 Describe how you would ensure *valid* results. To collect valid results, only the independent variable should be changed and you need to control any other variables that could affect the results. Think of other factors that could affect the rate of respiration and make sure that you have considered ways of controlling or monitoring these.

5 Link to the question stem to explain how you would *compare* the rates. To compare the rates of respiration in the mouse and woodlouse, you need take their differing size into account and convert the rate of oxygen uptake into a rate per gram.

6 Apply your scientific knowledge to *comment* on expected results and conclusions. This final step requires you to apply your knowledge of respiration to movement or the maintenance of a constant body temperature.

Allow yourself 10 minutes to answer this question.

Mark scheme

Modifications:

1 size of container (larger for mouse);
2 volume of potassium hydroxide (greater for mouse to absorb all carbon dioxide);

Ethical treatment:

3 animals removed from container promptly/gauze or mesh over potassium hydroxide to protect animals;

Collection of valid results:

4 temperature controlled (as it affects rate of respiration);
5 mass of animals found and rate calculated per gram;

Comment on results and conclusion:

6 rate of respiration higher in mouse because they have a higher metabolic rate/because they are endotherms/because they have a constant body temperature;

Sample answer

Place the woodlouse in the respirometer with potassium hydroxide to absorb the carbon dioxide. Note the start position of the drop of liquid and start the stop watch. Record the position of the drop of liquid after 10 minutes.

A larger container would be needed for the mouse. ✓ Using a larger container and a larger animal would require a larger volume of potassium hydroxide to absorb all of the carbon dioxide. ✓

Marking points 1 and 2 are awarded here for a clear description of the modifications that would be needed.

Handle animals gently and keep handling to a minimum to reduce distress. Remove animals from the respirometer as soon as the investigation is complete and return them to their containers or habitats. ✓ Cover the potassium hydroxide with gauze so that it does not come into contact with the animals as it can cause irritation or burns. ◄

The student has made two very good points relating to the ethical treatment of animals, but a maximum of 1 mark was available for this. Marking point 3 is awarded here.

Temperature will affect the rate of respiration, so place the respirometer in a water bath and monitor the temperature using a thermometer. ✓ Take repeats for each animal so that mean rates of oxygen uptake can be calculated.

Marking point 4 is awarded here for a very good description of how and why temperature should be controlled.

Calculate the distance moved by the drop of liquid in mm per minute. Find the mass of the mouse and the mass of the woodlouse, then calculate the distance moved by the liquid in mm per minute per gram. ✓ ◄

This is a very clear description of the results that will be collected, and marking point 5 is awarded.

This final sentence clearly links the expected result to a reason why a mouse would have a higher rate of respiration, for marking point 6.

The mouse would be expected to have a higher rate of respiration than the woodlouse as the mouse is a mammal and requires a high rate of respiration to maintain a constant body temperature. ✓

You should know

> You need to meet all of the CPAC criteria to pass your practical endorsement. Make sure that your practical work is of a high standard throughout the course and act on feedback from your teacher.

> You will be assessed on practical skills in the exams. Review these skills before your exams so that you can answer questions about procedures, results and conclusions.

> You may be assessed on your ability to modify or adapt procedures. Make sure that you are familiar with all of the apparatus and techniques used during your course.

> The exams will include unfamiliar practical procedures. Make sure that you can apply your knowledge to novel contexts.

> Some of the exam questions assessing practical skills require extended writing. Structure your answer logically and use any bullet points or suggested content given in the question.

5 Study and revision skills

Learning objectives

> To develop new techniques of remembering and understanding the specification content

> To use past paper questions and mark schemes effectively for exam preparation

> To analyse and interpret exam questions more accurately

> To evaluate answers to exam questions using both mark schemes and examiners' reports

This chapter will firstly focus on the core skills of remembering and understanding the specification content before suggesting ways in which you can develop the higher-order skills of analysing, applying, creating and evaluating. The skills that you will develop in this chapter are transferable skills that you will be able to apply to the study of your other A-level subjects, and hopefully at university.

Core study skills

The core skills of knowledge and understanding account for about a third of the overall A-level mark. Students often find biology A-level quite 'content heavy' because there is a lot to learn and a lot of facts to memorise. There is little call for your opinion when answering biology A-level questions — the upper chambers of the heart are called the atria, whether you think they should be or not!

Remembering

Remembering the specification is a fundamental step in preparing for the exam. You need to have a strong foundation of knowledge before you can access the higher-order skills of applying, evaluating and analysing.

Top tips

→ Revise regularly — the learning process should be ongoing and not just something to cram in at the end of the course. Read through your class notes on a daily or weekly basis, make your own revision notes, and revise thoroughly for end-of-topic tests. All of these processes will help you to store the key information in your long-term memory.

→ Learn and understand the 'core concepts' — these are the fundamental ideas that you study at the start of the A-level biology course and are the foundation on which the rest of the course is built, so it is really important that you learn and understand them.

Aiming for an A in A-level Biology

→ Plan your revision — you may want to make a revision timetable indicating what you will revise and when, but do not spend so long making the timetable that you run out of time to learn anything!

→ Refer to the specification — either print off a paper copy or download the PDF from the exam board website. Go through the specification and cross off or tick each topic when you have made your revision notes. Specifications tend to be very prescriptive and tell you exactly what you need to know. For example, AQA states that students learning about the roles of bacteria in the nitrogen cycle *do not* need to know the names of individual bacteria species, whereas OCR specifies that students need to know the roles of microorganisms, including *Nitrosomonas*, *Nitrobacter*, *Azotobacter* and *Rhizobium*. Learn the definitions of key terms, or structures of biological molecules, *exactly* as they appear in the specification.

→ Vary your activities — people learn in different ways and techniques that work for one student will not necessarily work for another. Using a range of activities will stop you from getting bored when revising. Plan a very *specific* activity rather than vaguely deciding to 'do some revision'. An example of this might be 'I am going to make notes about the structure and functions of triglycerides, and then answer the summary questions from page 32 of my textbook'.

Revision activities

Read through class notes, textbooks or revision guides

It is easy for your mind to wander while you are reading, so make sure that you *do* something with the information that you have just read. After reading a chapter, you could write a brief summary of what you have just read, cover the page and test yourself to see how much you remember, or test your recall using an online quiz. For example, type 'A-level biology lipids quiz' into a search engine and select one of the options.

> ## Take it further
> The Royal Society of Chemistry website has a section called 'Chemistry for Biologists', with notes and online multiple-choice tests on a range of topics including carbohydrates, lipids, enzymes, photosynthesis, respiration and energy flow.

Write summary notes from class notes or textbooks

Use the specification as guidance when you are writing your notes to avoid writing unnecessary content.

Make your notes once and do a good job of them — do not waste time rewriting them. Keep your notes together in an organised way.

Produce mind maps or spider diagrams

Drawing mind maps or spider diagrams encourages you to think more deeply about a topic because you have to decide how to organise the information. It can be useful to add colour and images to act as triggers when trying to recall information.

Produce flash cards

These are useful for learning definitions of key terms because you can write the key word on one side of the card and the definition on the other. Test yourself on both the key terms and definitions. If you get one incorrect, put it back in the pile so that you can test yourself on it again. Flash cards are handy because they can be carried around with you — you could test yourself while waiting for the bus, or during a car journey. As with revision notes, make sure that you have a method of keeping your cards together so that they do not get lost. Some flash cards are sold with a durable plastic cover and a metal ring to hold them together.

Prepare a revision wall

Display your mind maps, or write key terms on post-it notes. A brightly coloured and informative revision wall would help if you were trying to remember cycles or processes because you would see diagrams of them every day. Vary the information that you have displayed so that you do not get bored with it. You could look at the information while you are getting dressed, or brushing your hair!

Devise mnemonics or phrases

Sometimes you need to remember terms in a specific order, for example, the taxonomic hierarchy: **d**omain, **k**ingdom, **p**hylum, **c**lass, **o**rder, **f**amily, **g**enus and **s**pecies. A mnemonic such as 'dashing king Philip came over for great sushi' might help you to remember this sequence (you may find it easier to remember silly or funny phrases). The coenzyme NAD is involved in respiration, whereas NADP is involved in photosynthesis. You could remember this by linking the 'P' in NADP to 'p' for photosynthesis.

> ### Take it further
>
> Crash Course Biology is an educational YouTube channel with short videos (each about 10 minutes long) explaining specific topics — for example, the carbon cycle. The explanations are clear, make good use of animations, and are of a standard suitable for A-level. Watching and listening to these videos is a good way of supporting your learning and may provide you with another way of remembering the course content.

Understanding

Remembering the specification will not get you a grade A — you need to *understand* the content as well. Make sure that you understand a topic as soon as it is explained to you. Maximise your opportunities in the classroom by asking questions, and asking your teacher to explain something again if you do not understand it. Your school or college may have additional tutorial or study sessions that you can attend. Many students feel uncomfortable asking for help, or worry that they are going to ask a 'stupid' question. You could always ask your teacher a question at the end of the lesson when no-one else is listening. Remember that your teachers want to help you — that is why they teach.

Take it further

There are some excellent TED-Ed videos that will support your understanding of specific topics — for example, cell membranes, or the Calvin cycle. These professionally produced animations are about 5 minutes long and bring to life the words and ideas of educators. After watching the videos there are options to 'think' with a quiz on the video content, 'dig deeper' with further resources, and 'discuss' with open-ended questions or discussion points.

Exam questions are a very important resource when you are assessing your understanding of a topic. Your teacher may set you past paper questions for homework, or you may be finding past paper questions for yourself as part of your independent learning. When you are attempting these questions, you should never leave gaps. If you do not know the answer, then look it up. This is *active* learning because you are actively finding the information and selecting the appropriate pieces of information to answer the question. If you cannot find the answer, then write something sensible. You will never get a mark for a blank space.

As you become more experienced at answering past paper questions, you could try completing them under timed conditions. Remember that you will have approximately 1–1½ minutes per mark in the exam. After completing each question, self-assess your answer using the mark scheme. If you did not get the correct answer, or did not know how to answer the question, then look carefully at the mark scheme and check that you understand how to get the correct answer.

Peer mentoring is an excellent way to develop your understanding of a topic. If you are struggling to understand something, ask another student to explain it to you. They will probably explain it in a slightly different way from your teacher, and this may be enough to make things clear to you. If you *do* understand a topic really well, you could explain it to one of your peers. Thinking of ways to explain it and then verbalising this explanation will consolidate your own understanding.

Higher-order study skills

A student who has only focused on the core skills is unlikely to achieve higher than a grade D. This is based on the grade boundary for a grade D being about 30–35% of the total mark. Developing the higher-order skills is essential if you are aiming for an A.

Applying

Between 40% and 45% of the marks available at A-level will assess your ability to *apply* your knowledge. This involves you *using* your knowledge and understanding to answer questions on unfamiliar content.

Application of knowledge questions can be very off-putting because it may not be immediately obvious what the question is about. These questions may contain the names of specific enzymes, drugs or organisms that you have never heard of. The best way to prepare for these questions in the exam is by practising past papers.

✓ **Exam tip**

Be careful when using acronyms or abbreviations in the exam. Some are acceptable answers and will be included on the mark scheme, such as DNA, ATP and RuBisCO. Others may not be standard terms accepted by the examiner, for example SA for surface area, or RBC for red blood cell. If in doubt, write the word or phrase out fully in your answer and include the acronym or abbreviation in brackets afterwards — you can then use this in the rest of your answer.

✓ **Exam tip**

The new biology specifications were first examined in 2016, so there will not be many past papers to refer to. Remember to look at the legacy papers (examining the previous specification) as well. Much of the specification content will be the same and there will be similarities in the styles of exam question.

There is little point in answering past paper questions if you do not learn from your mistakes. Some students complete a past paper, mark it using the mark scheme and then check the grade boundaries. If they get the grade they wanted, then they are happy with their performance and move on to the next exam paper. Depending on the grade boundaries, it may be possible to achieve a grade A with 65% of the total marks, but this still means that you have lost *over a third* of the marks. Successful students learn from their mistakes so that they do not make the same mistake twice.

You could try a more targeted approach to answering exam questions:

1   Revise a specific topic, for example enzymes, and make sure that you can *remember* the key information about the mechanism of enzyme action, factors affecting the rate of enzyme-controlled reactions, and the role of inhibitors.

2   Check that you *understand* the topic by questioning yourself. For example: *Why* does temperature increase the rate of reaction? *How* do extremes of pH cause denaturation? *What* are the differences between competitive and non-competitive inhibition?

3   Now see if you can *apply* this knowledge and understanding by finding an exam question about enzymes. Most exam papers are available as PDF files and this will allow you to search the paper for key words rather than scrolling through the whole paper. Try searching for terms such as 'enzyme' or 'inhibitor'.

4   After answering *one* exam question on your chosen topic, mark your work using the mark scheme. If you lost any marks, then correct your work. What *should* you have written to get full marks?

5   Now find another exam question about the same topic to answer and self-assess. Hopefully, you will have learned from correcting the first exam question, so your second answer should be better.

This process will allow you to see if there are specific words or phrases that you need to include in your answers. Spending time on this focused approach to answering exam questions will teach you how to apply your knowledge to a range of contexts.

Evaluating

Analysing, evaluating and interpreting scientific information will account for 20–25% of the marks in your A-level exams. Throughout this book, you have been given guidance on evaluating data, written source material and practical procedures, but you can also develop your evaluative skills when studying.

When you are using mark schemes and assessing your responses to exam questions, be critical of your own answers. As well as referring to mark schemes, read the comments about questions in the examiners' reports. These comments are written by the principal examiners after they have seen hundreds of responses to the questions. The examiners' reports give you a valuable insight into the common errors and misconceptions that have caused students to lose marks.

 Exam tip

Examiners' reports are an underused resource that could make a big difference to your exam performance. The reports are written by the people who will actually mark your exam papers. Following their advice will help you to improve your answers and to write what the examiners want to read.

As well as marking your own work, try to mark other students' work. It is too easy to be nice to yourself when self-assessing your work because you know what you meant to write. You will be much more critical of other people's work. It is good to see how other students answer questions as it might give you new ideas.

Activity

Read the students' answers to the exam question and try to mark them using the mark scheme.

Explain how an increase in nitrate concentration in a lake can cause fish to die. (4)

Student A answer

The increased nitrate concentration causes an algal bloom on the surface of the lake. Submerged plants are unable to photosynthesise and eventually die. Decomposers break down the dead plant material and use dissolved oxygen as they respire. The lack of oxygen in the water causes fish and other aerobic organisms to die.

Student B answer

Nitrate causes algae to grow in the lake, preventing light from reaching aquatic plants. The plants cannot photosynthesise, so they die and are broken down by saprotrophic microorganisms. The death of the plants means that there is no food for the fish, and no oxygen is being produced by photosynthesis so the fish also die.

Mark scheme

1 (increase in nitrate concentration causes) increased growth of algae blocks light;
2 less/no photosynthesis so plants die;
3 broken down by saprobiontic bacteria (accept: saprotroph/saprophyte, neutral: decomposer);
4 (bacteria/saprobionts) use oxygen in respiration/respire aerobically so less oxygen for fish;

Feedback

Student A gains marking points 2 and 4. Point 1 has not been awarded because the answer does not refer to light being blocked. Point 3 has not been awarded because the mark scheme states that 'decomposer' is neutral (not an acceptable answer, but not incorrect). Student B gains marking points 2 and 3. Point 1 has not been awarded because there is no reference to an *increase* in algae (they were growing in the lake anyway). Point 4 has not been awarded because the lack of oxygen in the water has been related to the lack of photosynthesis by plants rather than to bacterial respiration.

Creating

Mind maps

Creating is a higher-order thinking skill that can be applied to your independent study. You could *create* mind maps as part of your revision schedule. Mind maps were referred to earlier as a method of remembering the content, but they are also very useful for creating synoptic links between different topics. The more able students do not memorise standalone facts, but demonstrate their ability to create links and relate these facts to each other.

Activity

Transport across membranes is an important concept that has links to several topics in the specification. Start in the middle of a page by writing 'Transport across membranes'. You may want to add a diagram at this point to show the fluid-mosaic structure of membranes and/or brief notes about membrane structure. Now add four main branches from your main title and label these branches 'Diffusion', 'Osmosis', 'Active transport' and 'Endocytosis/exocytosis'. For each of these four branches, write a definition of the term and draw diagrams to represent each process. Use colour to help you see the levels of organisation in your mind map.

We will now just focus on the 'Diffusion' branch: add two branches from diffusion and label these 'Simple diffusion' and 'Facilitated diffusion'. This is when you start looking for links across the specification — which other biological processes are dependent on diffusion?

- Large, multicellular organisms have specialised gas exchange surfaces to increase the rate of *diffusion* of oxygen and carbon dioxide.
- The ileum has adaptations to increase the rate of *diffusion* of the products of digestion into the bloodstream.
- Neurotransmitters move across synapses and neuromuscular junctions by *diffusion*.
- The synthesis of ATP by oxidative phosphorylation requires the *diffusion* of protons across the inner mitochondrial membrane through ATP synthase.
- The plant growth factor indoleacetic acid (IAA) moves towards the shaded side of a shoot by *diffusion* (and results in positive phototropism).

These are some examples that you may have covered during your A-level course. Once you start adding branches, you will be surprised how interlinked many topics are.

Complete your mind map by looking for links with the other mechanisms of transport across membranes.

Comparison tables

Another way of identifying links is by creating tables of comparison. There are lots of opportunities in biology to look for similarities and/or differences between molecules, processes or organisms. Table 5.1 is an example of a simple table comparing the structure and function of polysaccharides.

Table 5.1 The structure and function of polysaccharides

| | Polysaccharide | | | |
	Starch (amylose)	Starch (amylopectin)	Glycogen	Cellulose
Monomer	α-glucose	α-glucose	α-glucose	β-glucose
Bond between monomers	1,4 glycosidic	1,4 and 1,6 glycosidic	1,4 and 1,6 glycosidic	1,4 glycosidic
Chain structure	Coiled and unbranched	Long and branched	Short and highly branched	Long, straight and unbranched
Function	Storage (energy)	Storage (energy)	Storage (energy)	Structural

You could create tables or diagrams for the following suggestions to help you identify similarities and differences:

→ biological molecules (carbohydrates, proteins and lipids)
→ messenger RNA, transfer RNA and ribosomal RNA
→ Calvin cycle and Krebs cycle

Aiming for an A in A-level Biology

→ gas exchange surfaces in insects, fish, dicotyledonous plants and mammals

→ DNA replication, transcription and translation

→ hormonal and nervous coordination

Model answers

The use of exam questions and mark schemes has already been mentioned because it is an essential feature of exam success. These questions and mark schemes should be used in a variety of different ways to develop your exam technique. You could *create* model answers using the mark schemes for guidance. Biology mark schemes tend to be written as numbered points or bullet points rather than in continuous prose. You could look at these points and use them to write a 'perfect' paragraph that includes every marking point.

Activity

Look at the exam question and mark scheme. Use the bullet points to write a model answer and then compare it with the possible response shown.

Organic molecules are transported from the leaves of a plant to the roots. The mass flow theory suggests that organic molecules move from high pressure in the leaves to lower pressure in the roots. Describe how a high pressure is produced in the leaves. (4)

Mark scheme

- sucrose from companion cell to (phloem) sieve tube;
- by active transport/cotransport with hydrogen ions;
- lower/more negative water potential in sieve tube;
- water enters sieve tube by osmosis;

Possible response

Sucrose is actively transported ✓ from the companion cells to the sieve tube elements ✓. This reduces the water potential in the sieve tube element ✓ so water enters by osmosis ✓ creating a high hydrostatic pressure.

You could try to create your own exam question. This would be difficult to do, but you could base your question on a previous question. For example, if a previous essay question had asked you to describe the importance of proteins, or the importance of carbohydrates, then you could create a question that asked about the importance of water, or the importance of temperature.

Another approach that would encourage you to think more deeply about an exam question is to create your own mark scheme and then compare it with the actual mark scheme, to see if you included the correct marking points.

Analysing

Throughout this book there have been activities to develop your analytical skills in response to data, written source material and practical work. You can also put these skills to good use when analysing exam questions and whole question papers.

Before you can answer an exam question, you need to analyse it to make sense of the information you have been provided with and to determine what you are being asked to do.

→ Use a highlighter pen and identify the command word(s).

→ Refer to a glossary and check that you understand the meaning of the command word(s).

→ Look at any diagrams and read the labels carefully.

→ Look at any graphs or tables of information and identify patterns in the data.

→ Look at the mark allocation to give you an indication of the number of points you need to write.

As well as analysing individual exam questions, you should analyse whole exam papers to familiarise yourself with their structure and layout, and to look for patterns or trends in the content and style of questions.

> **Exam tip**
>
> Completing full exam papers under timed conditions as part of your exam preparation will help you to check that you can complete the paper in the allocated time. Put a clock or stop watch on the table in front of you so that you create the feeling of time pressure.

Activity

Make a spreadsheet or table with the specification reference in the first column and the years of the exam papers in the subsequent columns. Look through the exam papers for each year and tick the topics that have been covered in the paper. Analyse the table and look for patterns: are there topics that come up every year? Or is there a topic that has not been assessed recently?

Wellbeing

You need to look after your mental and physical health during this exam preparation period. It is not healthy to study all of the time and you need to remember the following:

→ Diet — eat a healthy, well-balanced diet with plenty of fruit and vegetables.

→ Relaxation — read a book, play a game, or whatever you need to do to wind down.

→ Exercise — get away from your desk and clear your head, even if you just go for a walk.

→ Sleep — you must get sufficient sleep, or you will not be able to process information effectively.

→ Socialising — make time to interact with friends and family to take your mind off your work.

The exam

On the day of the exam, you may feel nervous and worried. Make sure that you allow yourself plenty of time to get to the correct room. Try to have something to eat, and take a drink of water with you. Check that you have all of the equipment that you need:

→ Good-quality black pens (ballpoint or ink). Take at least two.

→ A scientific calculator that you know how to use. You may have to calculate standard deviations or use logarithmic functions.

→ Sharp pencils (and a rubber) for plotting graphs or drawing diagrams.

→ A ruler with mm divisions that you can use for accurate measuring, or for drawing *straight* lines.

When you are given the exam paper, remember that you do not have to answer questions in the order they are given. If you see a question on a topic you feel confident about, answer that question first. It will boost your confidence and make you more relaxed about the rest of the paper. If you encounter a question that you do not know how to answer, do not spend too much time on it. Move on to the next question and come back to it at the end if you have time.

Keep an eye on the time to make sure that you attempt all questions.

Be wary of changing answers unless you are certain that your first answer was wrong. Students often review their work and cross out their correct answer only to replace it with an incorrect answer!

> ✓ **Exam tip**
>
> Always use a pencil to plot a graph in the exam. The grid may be printed in the question booklet and you will not be able to get a new question booklet if you make a mistake.

You should know

> Make your revision an ongoing process throughout the course and explore new ways of learning.

> Remember that learning facts is not the ultimate goal. You need to push yourself continually to check your understanding and ability to apply your knowledge.

> Keep looking for possible links within or between topics, and new ways of grouping the content together.

> Use the exam papers, mark schemes and examiners' reports in a variety of ways to maximise your learning opportunities.

Exam board focus

Learning objectives

> To provide an overview of course content for each exam board

> To describe assessment by each exam board

> To list typical question stems that assess the higher-order skills

> To identify common ways in which students lose marks in exams

> To describe the exam techniques typically required by the top mark band

> To identify any unique requirements of each exam board

This chapter will provide you with a very brief overview of assessment by each exam board and will identify some of the key features displayed by an A-grade student. Although this chapter is designed to focus on the individual exam boards, the final 'Advice from examiners' section is relevant to all students.

The exam boards that will be discussed in more detail in this chapter are:

→ AQA

→ CCEA

→ Edexcel

→ Eduqas

→ OCR

→ WJEC

Activity

Search for your exam board and specification using a search engine, for example, 'OCR A-level Biology'. Download and/or print off a copy of the specification — this is the definitive guide to the course content and you will only be assessed on content from the specification. The website may have skills sheets, practical guidance and other useful resources that will support your independent learning.

AQA

Core content and assessment

The A-level course for AQA covers eight main topics assessed in three written papers. All three papers include structured short- and long-answer questions. Paper 1 has extended-response questions totalling 15 marks, and paper 2 has a 15-mark comprehension question. Paper 3 includes critical analysis of given experimental data for 15 marks and one 25-mark essay from a choice of two titles.

There are also 12 required practicals that must be completed as part of the course. The exams will assess your knowledge and understanding of the course content *including the required practicals*.

Assessment of higher-order skills

Typical exam questions assessing higher-order skills will expect you to demonstrate your ability to analyse, interpret and evaluate scientific information or evidence. Questions assessing these skills may have question stems that include the phrases listed below:

→ 'Explain the results...' (4)
→ 'Use Figure X and your knowledge of...to explain why...' (4)
→ 'Suggest an explanation for the results in Figure X.' (4)
→ 'Explain how these data support this suggestion.' (4)
→ 'Do the data in Table X support this conclusion? Give reasons for your answer.' (4)

Chapter 1 (Quantitative skills) helps you to prepare for this type of question, and chapter 3 (Writing skills) explains how to structure your answers.

The final question on paper 3 is a 25-mark essay on a major idea or theme. The essay assesses your knowledge, understanding and ability to apply information, but it also requires you to demonstrate your synoptic skills (use of information from across the specification). AQA is the one exam board that states in a mark scheme that you can only access the highest marks for the essay (24/25 or 25/25) if you show evidence of further reading or greater breadth of study.

Essay guidance from AQA examiners

→ An introduction and conclusion are not required.
→ Few candidates include content from beyond the specification, even though they could just use examples from their textbook to illustrate concepts.
→ If the essay begins with 'The importance of...', then to gain over 15 marks, all content must be linked to *why* it is important, for example, what would happen if the process did not occur?
→ The essay *must* be synoptic (drawing on several topic areas from across the specification) and it *must* be at A-level standard (no credit is given for GCSE material).
→ The essay needs to be well structured and clearly linked to the title; it is *not* 'Write everything you know about...'

Activity

Use a search engine to search for past papers for your exam board. Check that the papers are for the current specification. You should be able to access both the question papers and the mark schemes so that you can complete and self-assess papers before the actual exam. Familiarise yourself with the structure of the paper and make sure that you know how to respond to each type of question by focusing on command words.

CCEA

Core content and assessment

To achieve the full A-level qualification for CCEA, students need to complete both the AS units and the A2 units. The AS units make up 40% of the full A-level qualification, and the A2 units contribute 60%.

The external written examinations at both AS and A2 include structured questions and essay questions. The AS essay questions are worth 15 marks and the A2 essays are worth 18 marks.

Practical skills are assessed in external written examinations and teacher-marked practical assessments at both AS and A2.

Assessment of higher-order skills

Assessment at A2 will provide opportunities to demonstrate higher-order thinking skills by incorporating a wide range of question types, questions with an increased incline of difficulty and a decrease in structuring.

The high-achieving students are able to:

→ demonstrate a *thorough* knowledge and understanding of biological concepts and processes, and to present this information clearly using correct scientific terminology
→ accurately analyse and interpret complex data and write clear, logically structured arguments and evaluations
→ apply biological knowledge and understanding to both familiar and unfamiliar contexts

Questions assessing the higher-order skills will include evaluative tasks that may use the following phrases:

→ 'Using the information in the table, suggest...' (4)
→ 'Analyse the data in the table and account for the pattern shown.' (4)
→ 'Analyse the information provided and explain...' (4/5)
→ 'Evaluate the information provided and give a possible explanation for...' (3)

Refer to chapter 1 (Quantitative skills) for guidance on interpreting data and responding to this style of question.

Assessment of higher-order skills will also include synoptic questions that require you to make connections between different sections of the specification. The quality of your written communication will be assessed in extended-writing questions. A-grade students will demonstrate excellent quality of written communication and will be able to write clearly, logically and coherently, with correct use of specialist vocabulary. The extended-writing questions will be worth either 15 or 18 marks and may include command words and phrases such as:

→ 'Describe the similarities and differences between...' *and* 'Discuss how the different methods (of transport) are necessary to...'

→ 'Describe and explain the sequence of events...'

Chapter 3 (Writing skills) contains a lot of advice on planning and structuring answers to these essay-style questions.

Edexcel

Core content and assessment for Biology A (Salters-Nuffield)

The biology A-level for Edexcel Biology A (Salters-Nuffield) includes eight main topics assessed in three written papers. Papers 1 and 2 may include multiple-choice, short, open-response, calculation-based and extended-writing questions. Paper 3 has a section based on a pre-released scientific article and will include synoptic questions drawing on two or more different topics. Paper 3 also includes questions that target the conceptual and theoretical understanding of experimental methods.

There are 18 core practicals that must be completed during the course, and these will be assessed in the written papers.

Note that this is the only specification that includes the assessment of a scientific article pre-released 8 weeks before the examination. You must read this article carefully using the critical reading skills developed from chapter 2 (Reading skills).

Core content and assessment for Biology B

Edexcel Biology B has 10 main topics that are assessed across three written papers. Papers 1 and 2 may include multiple-choice, short, open-response, calculation-based and extended-writing questions. In addition to these types of question, paper 3 might include synoptic questions that draw on two or more different topics, and questions that target the conceptual and theoretical understanding of experimental methods.

There are 16 core practicals, which are mainly assessed in paper 3.

Assessment of higher-order skills

The Edexcel papers for both Biology A and Biology B have a lot of data analysis questions that require you to look for patterns in data and apply your biological knowledge to an explanation of these patterns. These questions typically begin with the phrase 'Analyse the data...' and are worth between 3 and 5 marks.

Your ability to apply knowledge and construct a well-sequenced response is assessed using extended-response questions worth a maximum of either 6 or 9 marks.

Extended-response questions may contain the following command words and phrases:

→ 'Discuss the possible consequences...'   (6)

→ 'Evaluate the use...'   (6)

→ 'Explain the effect...' (6)

→ 'Analyse the data...' (6)

→ 'Analyse the data and use your knowledge of...to evaluate the claim that...' (9)

Chapter 1 (Quantitative skills) provides guidance in the use of data when answering this type of question.

These questions are marked using level-of-response marking. In this type of marking, the answer is assigned to a mark band and then the specific mark within that mark band is determined. The top mark band (level 3) means that the student can be awarded 5 or 6 marks for a 6-mark question, or 7–9 marks for a 9-mark question. The key features of a level-3 (top mark band) answer are as follows:

→ Scientific reasoning is linked to evidence throughout the answer.

→ The answer is clear and logically structured.

→ Relevant material is selected and integrated into the response.

→ Evidence-based conclusions/decisions are made.

Practise planning and structuring your answers to these essay-style questions using the guidance in chapter 3 (Writing skills).

Eduqas

Core content and assessment

The A-level course includes three components and a 'Core concepts' unit. The three components are assessed in three written papers, and content from the Core concepts unit is assessed in *all three* written papers. Students also study *one* of three options and answer questions on this option in section B of the Component 3 paper. All three papers include a range of short and longer structured, compulsory questions.

Each component includes specified practical work to be assessed in the written papers.

Assessment of higher-order skills

The Eduqas exam papers include a lot of data analysis and application of knowledge questions that are phrased in some of the following ways:

→ 'Based on these results, identify... Explain how you reached these conclusions.' (5)

→ 'Using the information provided, explain why...' (5)

→ 'With reference to the information provided, explain...' (4)

→ 'Analyse the data given in the table and draw conclusions... Explain how you reached these conclusions.' (3)

→ 'Predict how... Explain the basis for your prediction.' (5)

→ 'Using the apparatus shown, design a method to investigate...' (5)

→ 'Using information in the diagram and the table, what conclusion can be made about... Explain your answer.' (4)

There are a lot of marks available for this style of question. Use chapter 1 (Quantitative skills) for guidance on how to approach them.

The final question on each the three papers is a 9-mark extended-response question. These questions may provide you with data in various formats and then ask you to do the following:

→ 'Using the information given, determine how... Explain how you arrived at your conclusion...'
→ 'Explain the ways that...'
→ 'Suggest the effect...'

The quality of extended response (QER) will be assessed using a level-of-response style of marking, where your answer is assigned to a mark band and then the mark within that mark band is determined.

To achieve the top mark band (7–9 marks), your answer needs to meet the following criteria:

→ Includes a full account with *detailed* explanations of scientific processes.
→ Answer is *articulate* with accurate use of scientific vocabulary.
→ The account shows *sequential reasoning* with correctly linked and relevant points.
→ There are no errors or significant omissions.
→ Correct reference is made to data in support of the explanation.

Chapter 3 (Writing skills) provides guidance on how to structure these essay-style exam questions to maximise your mark.

OCR

Core content and assessment

The content for OCR is split into six teaching modules assessed in three written papers. The written papers include multiple-choice questions, short-answer questions (structured questions, problem-solving, calculations and practical) and extended-response questions.

There are also 12 practical activity groups (PAGs) that are assessed in all three of the written papers.

Assessment of higher-order skills

There are problem-solving questions on all three of the written papers that provide data and require you to analyse and interpret the data when answering the question. This type of question is assessing higher-order skills and may be phrased in one of the following ways:

→ 'Using the data and the information given, deduce a possible mechanism to account for...' (3)
→ 'Analyse the data and draw conclusions about...' (3)
→ 'Discuss the implications of...' (4)
→ 'Discuss the validity of the conclusions...' (3)

Refer to chapter 1 (Quantitative skills) for guidance on handling the data and worked examples that show how you could structure your answers.

Each paper includes extended-response questions worth 6 or 9 marks, which are marked using level-of-response marking. These questions often provide information in the form of tables, graphs or diagrams and require you to use these data in your response. Questions may be phrased in the following ways:

→ 'Using the information given, deduce why and how...' (6)

→ 'With reference to...explain...' (6)

→ 'Explain why these concerns are justified and suggest improvements to the investigation.' (6)

→ 'Comment on the results you might expect from this experiment and the conclusions you might draw.' (6)

→ 'Suggest how... Use the information given and your knowledge of...' (9)

A student's response is allocated to a mark band and then the mark within that mark band is decided. The top mark band (level 3) is for answers that would gain 5–6 marks out of 6, or 7–9 marks out of 9. An answer is deemed to be in the top mark band if it demonstrates these features:

→ Evidence is interpreted coherently.

→ Explanation is clear and thorough (all scientific content is addressed).

→ Ideas are clearly linked.

→ Good understanding of scientific principles is shown.

→ Data are critically evaluated.

→ Conclusions are reasoned.

→ The answer demonstrates a holistic grasp of theories.

→ The line of reasoning is well developed.

→ The answer is clear and logically structured.

→ All content is relevant.

→ Information is substantiated (supported by evidence).

The answer may not need to demonstrate all of the features listed, only those that are relevant. An A-grade student would be expected to write a factually correct account, as any errors would detract from the quality of the response. The answer should be well communicated and show clarity — for example, there should be no ambiguous statements or vague references to 'it'. The whole response should flow and read well; it should be well structured and easy to follow. All evidence should be used and any evaluation of conclusions should both support *and* challenge the conclusion.

Refer to chapter 3 (Writing skills) for guidance on how to approach these essay-style questions.

WJEC

Core content and assessment

The A-level qualification for WJEC includes two AS units and three A2 units. There are four written papers and a practical examination.

The four written papers all include a range of short and longer structured questions and an extended-response question.

There is also specified practical work that must be completed during the course to allow you to develop the knowledge and skills that will be assessed in the written papers.

Note that this is the only exam board that has a practical exam. The practical exam includes a 2-hour experimental task worth 20 marks, during which you will be given a set of apparatus and an examination paper containing the method to follow. There will then be a 1-hour practical analysis task for which you will be provided with experimental data to analyse.

Assessment of higher-order skills

There is a focus on the higher-order thinking skills of evaluation and justification in the exam papers for WJEC. These questions typically include phrases such as:

→ 'Evaluate this statement...'   (4)

→ 'Evaluate the validity of this statement.'   (4)

→ 'Evaluate the strength of their evidence and the validity of their conclusion.'   (4)

→ 'Using your knowledge of...justify this theory.'   (4)

You need to remember that this type of question needs a balanced response that considers the evidence for and the evidence against. You should also refer to chapter 1 (Quantitative skills) for guidance on data analysis.

Each of the written papers includes a quality of extended response (QER) question worth 9 marks. This type of question generally provides you with data in the form of tables, graphs and diagrams, and then requires you to refer to the data in your response. Typical wording for these QER questions might be as follows:

→ 'Using all of the information and your knowledge of...explain...' (9)

→ 'With reference to the information provided and your knowledge of...explain...'   (9)

To reach the top mark band in these level-of-response questions, your answer should include these features:

→ The account is articulate.

→ Reasoning is sequential (follows a logical structure).

→ The question is answered fully.

→ All material is relevant.

→ There are no significant omissions.

→ Scientific vocabulary is accurately used throughout.

It is important to plan and structure your answer carefully, so refer to chapter 3 (Writing skills) for guidance on how to approach these essay-style questions.

Advice for students — the ten golden rules

This advice is relevant to all students. It is based on the examiners' reports produced by all of the exam boards.

1 Read the question

→ Try not to rush.

→ Slow down, take your time and read the question carefully.

2 Follow the instructions in the question

→ Do what the question tells you to do and do not just repeat the stem of the question.

→ Instructions may tell you to 'give your answer *to the nearest whole number*' or 'give your answer *to the nearest minute*'. You will not get the mark if you do not do this.

→ Only include information that the question asks for:

  → If a question asks about inspiration, do not write about expiration.

  → If you are asked to write about sympatric speciation, do not describe allopatric speciation, because this would be a waste of time and space.

  → If you are asked for a feature of the cell *membrane*, it would be irrelevant to discuss other features of cells or cell organelles.

→ Give the number of responses required by the question — for example, 'suggest *two* reasons...'.

→ Do not repeat examples given in the question — for example, 'give *two other* adaptations...'.

3 Respond correctly to command words

→ Familiarise yourself with the command words and their meanings.

→ You must do what the command words ask you to do, or you will not get the marks. If asked to *explain* a graph, no marks are available if you *describe* the graph. If you are given data and asked to write an explanation, *use* the data to support your explanation, but no marks are available for *describing* the data.

→ Use the command word to focus your answer so that it does not contain irrelevant material.

4 Present your work clearly

→ Try to write as neatly and clearly as possible; the examiners can only mark what they can read.

→ Use black ink or a black ballpoint pen because your exam paper will be scanned and this will give a clear image.

→ Do not overwrite a mistake. Cross it out and rewrite it.

→ Write a clear, concise answer that fits into the space provided without using arrows, asterisks and answers scattered throughout the booklet.

→ There should be plenty of space provided for your response. Use the size of the space given as an indication of the length of the answer.

5 Use the correct scientific terminology

→ Take care with spellings of key terms. For example:

  → *lymphocyte*, not 'lympocyte'

  → *clonal* expansion (of B lymphocytes), not 'colonial' expansion

→ Learn the difference between similar-sounding words such as antibody, antigen and antibiotic.

→ Use comparative terms when necessary, for example, high*er* or *more*.

→ Avoid poor use of language, for example use:
  → *concentration* or *volume* rather than 'amount' or 'level'
  → *nerve impulse* rather than 'message' or 'signal'
  → *hydrolysis* rather than 'breakdown'
→ Be specific:
  → name a *specific* enzyme, for example DNA polymerase, rather than writing 'the enzyme'
  → name a *specific* bond, for example phosphodiester, not just 'a bond'

6 Use the information provided

→ Application questions may provide specific information in the form of a written introduction, photo, table or graph.
→ Some answers require you to use the stem to *inform* your answer, making it a specific application of your knowledge. Sometimes information in the stem of the question will help you with your answer.
→ If asked to use data, your answer must include reference to the data.
→ When you are analysing data, make it clear which piece of information you are referring to. Use the evidence that you are given and link any trends to your conclusions.

7 Structure your answer clearly

→ Essays and extended-response questions require you to use your own knowledge and understanding (with high levels of detail) together with any information provided in the stem of the question.
→ The best answers are concise and well written.
→ Plan your answer so that it is not rambling and unfocused.
→ Sequence your answer logically.
→ Make sure that all of the content is relevant and linked directly to the question.
→ Remember that the question is not 'write everything you know about…'. Your response must *answer* the question and should not be you 'dumping your brains' in response to a specific word!
→ AQA, CCEA and Eduqas require you to make synoptic links across several areas of the specification when answering the essay questions.

8 Practise the mathematical skills

→ A-level students should not lose marks because of a lack of basic mathematical skills.
→ Develop your mathematical skills throughout the A-level course.
→ Remember that 10% of the marks will be from the assessment of mathematical skills. You should be able to confidently use all of the mathematical skills, including:
  → calculating percentage change
  → displaying numbers in standard form and to an appropriate number of significant figures
  → interconverting units, especially mm, μm and nm, as these are the most commonly used units of measurement when referring to cells, cell structures and microorganisms

➜ using ratios (students tend to perform badly in questions using ratios)

➜ identifying uncertainties

➜ accurately reading values from graphs

➜ Show your working clearly, in logical steps.

➜ Give your answer to the same number of decimal places as those in the question, unless told otherwise.

9 Revise your practical skills

➜ Remember that the practical skills will be assessed on the written papers, so you must revise this content.

➜ You should be able to demonstrate and apply knowledge and skills gained from doing the practical work.

10 Apply your knowledge to the context of the question

➜ Many questions require you to *apply* your knowledge, so rote-learnt answers will not get credit.

➜ Avoid vague, generic responses.

➜ Your answer must be related to the context of the question. For example, natural selection or speciation should be directly related to the example of the organism given in the question.

Activity

Once you have completed and self-assessed a past paper, find the examiners' report for the exam paper. This report is written by the principal examiner at the end of the marking session and it contains very useful information about common mistakes and the specific skills or topics that students struggled with. Use this report to inform your revision and to develop your exam technique. By knowing the ways in which students tend to lose marks, you can learn from the mistakes of others.

You should know

> **Don't forget the practicals! You need to review both the theory and the skills because the practical assessment is worth at least 15% of the final mark.**

> **You do not need to get full marks to get grade A. About 70% of the raw mark should get you a grade A. This means that you can lose nearly a third of the marks and still get an A, so do not panic if you cannot answer a question!**

> **Maximise your marks by identifying your strongest skills and answering those questions first.**

> **Learn what each command word means so that you can focus your answer.**

> **Write clear, concise answers as neatly as you can and do not waste time writing long answers with irrelevant material.**

> **Remember your mathematical skills. You need to practise these skills because they account for at least 10% of the final mark.**